CONTEMPORARY

NUMBER POWER

A REAL WORLD APPROACH TO MATH

Addition, Subtraction, Multiplication, and Division

JERRY HOWETT

Bothell, WA • Chicago, IL • Columbus, OH • New York, NY

www.mheonline.com

Cover 5 119 Travelif/iStock Exclusive/Getty Images

Send all inquiries to:
Contemporary/McGraw-Hill
130 E. Randolph St., Suite 400
Chicago, IL 60601

ISBN: 978-0-07-657794-1
MHID: 0-07-657794-5

Printed in the United States of America.

6 7 8 9 10 RHR 17 16 15

The McGraw·Hill Companies

TABLE OF CONTENTS

Multiplication

Division

Posttest A

TO THE STUDENT

This Number Power book is designed to help you build and use the basic whole number math skills found in the workplace and in everyday experiences.

The first section of the book, called Building Number Power, provides step-by-step instruction and plenty of practice in addition, subtraction, multiplication, and division of whole numbers. Work in these chapters begins with a skills inventory to pinpoint your strengths and weaknesses. Each chapter ends with a cumulative review to measure your progress and to show you which areas need additional work.

Because money is an important part of everyday life, some problems include numbers with dollars and cents. In our money system, whole dollar amounts are written to the left of the decimal point, and cents are written to the right.

The section called Using Number Power will give you a chance to put your skills to work by applying them in real-life problems.

Learning how and when to use a calculator is an important skill to develop. In real life, problem solving often involves the smart use of a calculator—especially when working with large numbers. You need to know when an answer on a calculator makes sense, so be sure to have an estimated answer in mind. Remember that a calculator can help you only if you set up the problem and use the calculator correctly.

At the back of this book is an answer key for all the exercises and tests. Checking the answer key after you have worked through a lesson will help you measure your progress. Inside the back cover is a chart to help you keep track of your score on each exercise. Also included in the back are a list of measurements and a glossary of the mathematical terms used in this book.

The whole number skills that you learn, practice, and apply in this book are the building blocks for all of mathematics. Work your way slowly and carefully through this book to build a solid foundation for your increasing Number Power.

Pretest

This test will tell you which sections of the book you need to concentrate on. Do every problem that you can. Correct answers are listed by page number at the back of the book. After you check your answers, the chart at the end of the test will guide you to the pages of the book where you need work.

1. Which digit in the number 27,346 is in the *hundreds* place?

2. The area of Greece is 50,949 square miles. Round the area to the nearest *thousand* square miles.

3. Which digit in the number 2,476,500 is in the *ten thousands* place?

4. The town of Hudson Village has 3,468 registered voters. What is the number of registered voters rounded to the nearest hundred?

5.
$$\begin{array}{r} 46 \\ +\ 52 \\ \hline \end{array}$$

6.
$$\begin{array}{r} 908 \\ 3{,}470 \\ +\quad 65 \\ \hline \end{array}$$

7. $3 + 26 + 47 + 4 =$

8. $21{,}859 + 9{,}786 =$

9. $258{,}497 + 87{,}063 =$

10. Round 923 to the nearest *hundred.* Round 8,609 to the nearest *thousand.* Round 38,519 to the nearest *ten thousand.* Then add the rounded numbers.

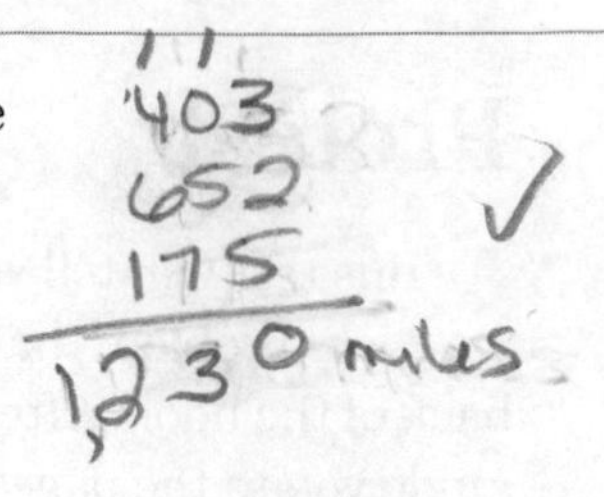

11. The driving distance from Los Angeles to San Francisco is 403 miles. The distance from San Francisco to Portland, Oregon, is 652 miles. The distance from Portland to Seattle is 175 miles. Find the total distance from Los Angeles to Seattle by way of San Francisco and Portland.

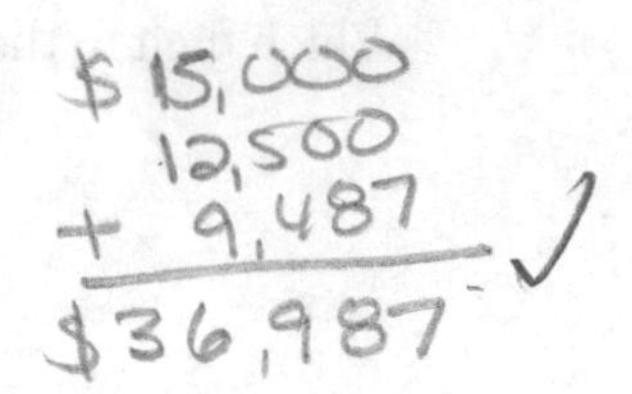

12. A concerned citizens group received a $15,000 grant from the federal government to improve a playground. They also received a $12,500 grant from the state and $9,487 in private donations. Find the total amount of funds the group has raised for playground improvements. $36,987

13. $\begin{array}{r} 167 \\ -\ \ 52 \\ \hline \end{array}$ 115

14. $\begin{array}{r} 6,324 \\ -\ 4,295 \\ \hline \end{array}$ 2,029

15. 1,827 − 959 = 868

16. 20,000 − 6,123 = 13,877

17. 803,400 − 29,753 = 773,647

18. Round each number to the nearest *hundred.* Then subtract the rounded numbers.

2,981 − 746

3,000 − 700 = 2,300

19. In a recent year, the average price for an existing single-family house in El Paso was $131,800. The average price for an existing single-family house in Dallas was $150,700. How much more was the average price of an existing house in Dallas than the price of an existing house in El Paso?

$150,700 − $131,800 = $18,900 Average price

20. In a 10-year period, the number of home healthcare workers in the United States increased from 867,100 to 1,307,000. By how many workers did the number increase?

1,307,000 − 867,100 = 2,439,900

21. $\begin{array}{r} 54 \\ \times\ 21 \\ \hline \end{array}$

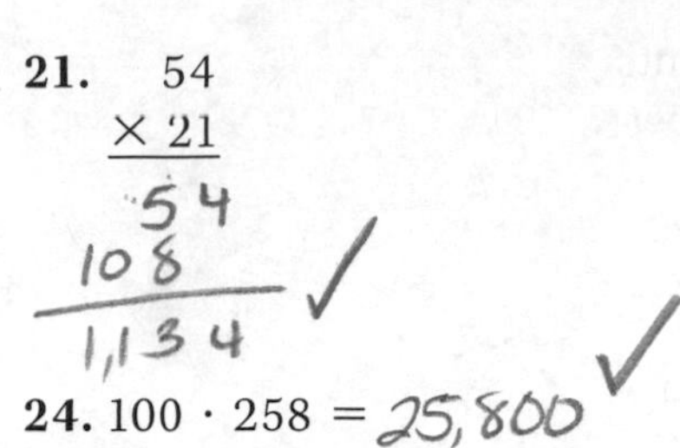

22. $\begin{array}{r} 358 \\ \times\ \ 9 \\ \hline \end{array}$

23. $(76)(54) =$

24. $100 \cdot 258 =$

25. $37(1{,}925) =$

26. Round 82 to the nearest *ten.* Round 9,271 to the nearest *thousand.* Then find the product of the rounded numbers.

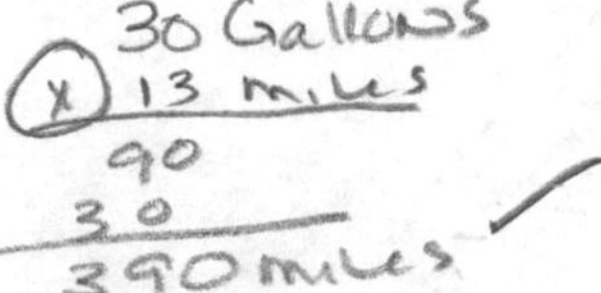

27. Steve's sport-utility vehicle gets 13 miles on 1 gallon of gasoline in the city. The gasoline tank holds 30 gallons. How many miles of city driving can Steve get on a full tank of gasoline?

28. Sound travels at a speed of 1,128 feet per second. Find, to the nearest *hundred* feet, the distance a thunderbolt travels in 20 seconds.

29. $8\overline{)\,536}$

30. $7\overline{)\,5{,}645}$

31. $1{,}288 \div 23 =$

32. $\dfrac{51{,}100}{70} =$

33. $\dfrac{8{,}313}{163} =$

34. Find the answer to the problem $69{,}736 \div 92$ to the nearest *ten.*

35. Diana works for a newspaper delivery service. Sunday papers are packaged together with 15 papers in each bundle. How many bundles are required to package the 3,000 papers that are delivered each Sunday to the suburb of Grey Gardens?

36. A leading player in the National Basketball Association scored 1,886 points during the season. He played a total of 82 games. Find the average number of points he scored per game.

PRETEST CHART

If you miss more than one problem in any section of this test, you should complete the lessons on the practice pages indicated on this chart. If you miss only one problem in a section of this test, you may not need further study in that chapter. However, before you skip those lessons, we recommend that you complete the review test at the end of that chapter. For example, if you miss one addition problem, you should pass the Addition Review (pages 29–30) before beginning the chapter on subtraction. This longer inventory will be a more precise indicator of your skill level.

Problem Numbers	Skill Area	Practice Pages
1, 2, 3, 4	place value	6–12
5, 6, 7, 8, 9, 10, 11, 12	addition	15–28
13, 14, 15, 16, 17, 18, 19, 20	subtraction	33–50
21, 22, 23, 24, 25, 26, 27, 28	multiplication	56–78
29, 30, 31, 32, 33, 34, 35, 36	division	84–110

BUILDING NUMBER POWER

PLACE VALUE

Understanding Place Value

What do these four numbers have in common?

4,321 1,234 3,412 2,143

You probably noticed that they are all four-**digit** numbers, but did you notice that all four numbers are made up of the same digits: 1, 2, 3, and 4? The digits are the same, but each number has a different value. This is because the digits are in different **places** in each number. In our number system the place of the digit tells you its value. In other words, each digit in a number has a **place value.**

In the box at the right, the four numbers are written with each digit under the name of the place in which it stands.

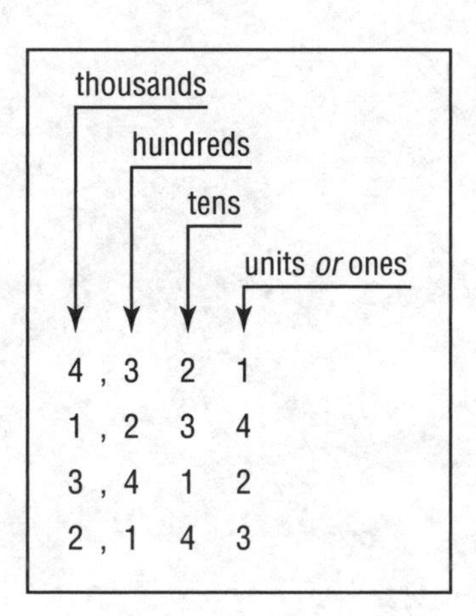

EXAMPLE **What is the value of the digit 4 in each of the numbers at the top of this page?**

The digit 4 in 4,321 is in the thousands place.
It has a value of 4,000.

The digit 4 in 1,234 is in the units or ones place.
It has a value of 4.

The digit 4 in 3,412 is in the hundreds place.
It has a value of 400.

The digit 4 in 2,143 is in the tens place.
It has a value of 40.

Use the number 7,856 to answer questions 1 to 6.

1. Which digit is in the thousands place?

2. Which digit is in the hundreds place?

3. What is the value of the digit 7?

4. What is the value of the digit 8?

5. What is the value of the digit 5?

6. What is the value of the digit 6?

Use the number 9,204 to answer questions 7 to 12.

7. Which digit is in the hundreds place?

2

8. Which digit is in the units place?

4

9. What is the value of the digit 9?

9,000

10. What is the value of the digit 2?

200

11. What is the value of the digit 0?

0

12. What is the value of the digit 4?

4

There are 5,280 feet in 1 mile. Use the number 5,280 to answer questions 13 to 16.

13. What is the value of the digit 5? 5,000

14. Which digit is in the hundreds place? 2

15. Which digit is in the tens place? 8

16. Which digit is in the units place? 0

Serena paid $6,750 for a used car. Use the number 6,750 to answer questions 17 to 19.

17. What is the value of the digit 7? 700

18. Which digit is in the thousands place? 6

19. Which digit is in the units place? 0

Mr. Johnson was born in 1924. Use the number 1924 to answer questions 20 to 22.

20. What is the value of the digit 2? 20

21. Which digit is in the thousands place? 1

22. Which digit is in the hundreds place? 9

The cost for each pupil at Greendale School is $8,459. Use the number 8,459 to answer questions 23 to 25.

23. What is the value of the digit 4?

400 hundreds

24. Which digit is in the thousands place?

8,000

25. Which digit is in the units place?

9

In his professional baseball career, Hank Aaron played in 3,298 games. Use the number 3,298 to answer questions 26 to 29.

26. What is the value of the digit 2?

200

27. Which digit is in the thousands place?

3

28. Which digit is in the tens place?

9

29. Which digit is in the units place?

8

30. Which of the following numbers has a 5 in the hundreds place?

a. 6,235

b. 5,623

c. 3,265

d. 2,536

Zeros and Place Value

What do these three numbers have in common?

6,200   6,020   6,002

You probably noticed that each number has four digits and also that each number is written with the digits 0, 2, and 6. The numbers are different because zeros hold different places in each number.

In 6,200 zeros hold the tens and units places.

1. In 6,020 zeros hold the ___Hundreds___ and ___Units___ places.

2. In 6,002 zeros hold the ___Hundreds___ and ___tens___ places.

3. In 9,000 zeros hold the ___Thousands___, ___tens___, and ___Units___ places.

4. In $400 zeros hold the ___tens___ and ___Units___ places.

The chart below lists the first seven places in the whole number system. Below the place names are three examples of numbers written with the digits 0, 1, 2, 3, 4, 5, 6, 7, 8, and 9.

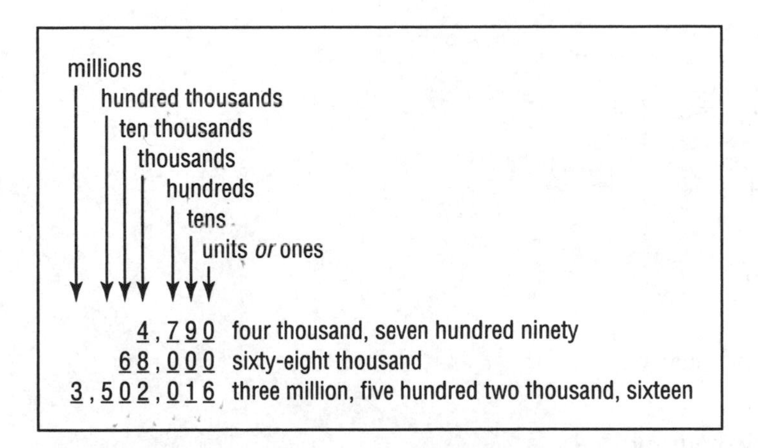

In 4,790 notice the zero. The zero holds the units or ones place. Notice also that commas separate the places in groups of three starting from the right. The number 4,790 is the correct way to write *four thousand seven hundred ninety.*

The number 68,000 has three zeros. The zeros hold the hundreds, tens, and units places.

The number 3,502,016 has two zeros. The zero between the 5 and the 2 holds the ten thousands place. The zero between the 2 and the 1 holds the hundreds place.

For problems 5 to 10 circle the number that is written correctly.

5. five thousand six hundred 5,060 5,600 5,006

6. seven thousand two 7,200 7,020 7,002

7. forty-four thousand, nine hundred 44,009 44,900 44,090

8. sixty-two thousand, four hundred three 62,403 60,243 62,430

9. eight hundred twenty thousand, seven hundred 827,000 820,007 820,700

10. two million, three hundred thousand 2,300,000 2,000,300 2,003,000

Write the following numbers.

11. forty-nine thousand, seven hundred thirty-six 49,736

12. four thousand, eighty-two 4,082

13. two hundred four thousand, nine hundred seventy-two 204,972

14. three hundred thirty-nine 339

15. five thousand, one hundred nine 5,109

16. thirteen thousand 13,000

17. eight hundred two 802

18. thirty-four thousand, eighty-six 34,086

19. six hundred eighty thousand 680,000

20. two hundred forty thousand, seven hundred 240,700

21. one million, three hundred forty thousand, six hundred 1,340,600

22. three million, four hundred twenty-five thousand, one hundred 3,425,100

23. nine million, eight hundred seventy-three thousand 9,873,000

24. three hundred thirty-three thousand, three hundred 333,300

Rounding Whole Numbers

A **round number**—a number that ends with zeros—is easy to work with. Later, when you learn to estimate answers, you will often round the numbers in problems to similar numbers that end with zeros.

Is the number 26 closer to 20 or to 30? The number 26 is closer to 30 than to 20. The number 26 rounded to the nearest ten is 30.

To round a whole number, do the following steps.

STEP 1 Underline the digit in the place you are rounding to.

STEP 2 **a.** If the digit to the right of the underlined digit is *greater than or equal to 5,* add 1 to the underlined digit.

b. If the digit to the right of the underlined digit is *less than 5,* leave the underlined digit as it is.

STEP 3 Change all the digits to the right of the underlined digit to zeros.

EXAMPLE 1 **Round 479 to the nearest ten.**

STEP 1 Underline 7, the digit in the tens place. 4$\underline{7}$9 → 480

STEP 2 The digit to the right of 7 is 9. Since 9 is greater than 5, add 1 to 7. 7 + 1 = 8

STEP 3 Change the digit to the right of 8 to 0.

EXAMPLE 2 **Round 4,316 to the nearest hundred.**

STEP 1 Underline 3, the digit in the hundreds place. 4,$\underline{3}$16 → 4,300

STEP 2 The digit to the right of 3 is 1. Since 1 is less than 5, leave 3 as it is.

STEP 3 Change the digits to the right of 3 to zeros.

Round each number to the nearest ten.

1. 76 **2.** 128 **3.** 261

4. 5,069 **5.** 288 **6.** 92

7. 14,355 **8.** 334 **9.** 983

Round each number to the nearest hundred.

10. 294 – 300

11. 3,528 – 4,000

12. 661 – 700

13. 2,319 – 2,000

14. 453 – 500

15. 18,766 – 18,800

16. 927 – 900

17. 6,859 – 7,000

18. 371 – 400

Round each number to the nearest thousand.

19. 3,614 – 400

20. 8,197 – 8,000

21. 12,543 – 13,000

22. 8,469 – 8,000

23. 235,520 – 240,000

24. 9,230 – 9,000

25. 26,612 – 30,000

26. 1,199 – 1,000

27. 62,903 – 63,000

Round each number to the nearest ten thousand.

28. 14,500 – 15,000

29. 26,125 – 30,000

30. 142,890 – 143,000

31. 81,777 – 82,000

32. 249,000 – 200,000

33. 173,208 – 170,000

34. 526,400 – 530,000

35. 78,312 – 70,000

36. 1,237,000 – 1,000,000

37. Sam and Serena bought a new car for $22,942. Which of the following rounds the price of the car to the nearest thousand dollars?

a. $22,940
b. $22,900
c. $23,000
d. $20,000

38. A real estate appraiser told Mrs. Vega that the value of her house was $136,500. Which of the following expresses the value of the house to the nearest ten thousand dollars?

a. $100,000
b. $136,500
c. $137,000
d. $140,000

39. Cal Hamilton ran for state assembly in his district. He won with a total of 213,802 votes. Which of the following expresses the number of votes to the nearest ten thousand?

a. 200,000
b. 210,000
c. 213,800
d. 214,000

ADDITION

Addition Skills Inventory

Do all the problems that you can. There is no time limit. Work carefully and check your answers, but do not use outside help. Correct answers are listed by page number at the back of the book.

1. $\begin{array}{r} 4{,}825 \\ +\ 2{,}164 \\ \hline \end{array}$

2. $\begin{array}{r} 51{,}937 \\ +\ 35{,}042 \\ \hline \end{array}$

3. $\begin{array}{r} 210{,}976 \\ +\ 384{,}012 \\ \hline \end{array}$

4. $\begin{array}{r} 23 \\ 54 \\ +\ 69 \\ \hline \end{array}$

5. $\begin{array}{r} 72 \\ 83 \\ 46 \\ +\ 54 \\ \hline \end{array}$

6. $\begin{array}{r} 28 \\ 50 \\ 63 \\ 48 \\ +\ 52 \\ \hline \end{array}$

7. $\begin{array}{r} 927 \\ 48 \\ 413 \\ +\ 52 \\ \hline \end{array}$

8. $\begin{array}{r} 46 \\ 519 \\ 53 \\ +\ 246 \\ \hline \end{array}$

9. $\begin{array}{r} 506 \\ 49 \\ 732 \\ +\ 88 \\ \hline \end{array}$

10. $41 + 638 + 23 =$

11. $215 + 23 + 44 + 7 =$

12. $8{,}126 + 75{,}634 + 29 =$

13. $52 + 3{,}497 + 8 + 21{,}046 =$

14. $476{,}279 + 861{,}557 =$

15. $16 + 38 + 4 + 125 + 2 =$

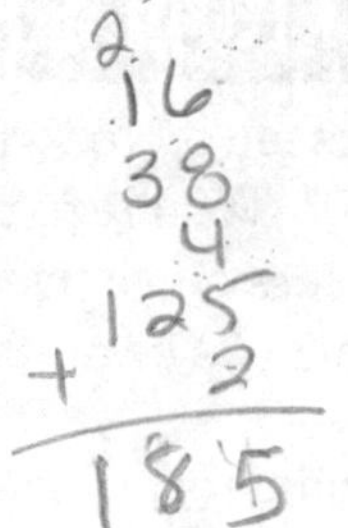

16. Round each number to the nearest *hundred* and add.

$6{,}954 + 1{,}326 + 4{,}579 \approx$

17. Round each number to the nearest *thousand* and add.

$29{,}847 + 31{,}866 + 49{,}230 \approx$

18. Round each number to the nearest *thousand* and add.

59,146 + 28,759 + 61,238 + 49,852 ≈

19. Round each number to the nearest *hundred* and add.

5,418 + 6,059 + 7,837 + 2,364 ≈

20. During a basketball season, the Atlanta Hawks won 56 games and lost 26. What total number of games did they play?

21. In an election for governor of Wyoming, the winner received 97,299 votes. The second-place candidate received 70,661 votes. The third-place candidate received 6,897 votes. Find the combined number of votes for the three leading candidates.

22. The attendance for 4 days at a county fair were 6,380 the first day, 5,963 the second day, 6,754 the third day, and 7,018 the last day. Round each day's attendance to the nearest hundred. Then find an estimate of the total attendance using the rounded numbers.

23. The following deductions were taken from Mr. Munro's paycheck: $15.86 for federal tax, $26.46 for social security, $8.36 for state tax, and $3.81 for city tax. Find the total amount of the deductions from Mr. Munro's check.

ADDITION SKILLS INVENTORY CHART

If you missed more than one problem on any group below, work through the practice pages for that group. Then redo the problems you got wrong on the Addition Skills Inventory. If you had a passing score on all five groups of problems, redo any problem you missed and begin the Subtraction Skills Inventory on page 31.

Problem Numbers	Skill Area	Practice Pages
1, 2, 3	addition facts	15–17
4, 5, 6, 7, 8, 9	adding and carrying	18–20
10, 11, 12, 13, 14, 15	adding horizontally	21
16, 17, 18, 19	rounding and estimating	23–24
20, 21, 22, 23	applying addition	25–28

Basic Addition Facts

This page and the following page will help you learn the basic addition facts. The addition facts are basic building blocks for further study of mathematics. You must *memorize* these facts. Check your answers to this exercise. Practice the facts you got wrong. Then try these problems again until you can get all the answers quickly and accurately.

Parts of an Addition Problem

4	← addend
+ 5	← addend
9	← sum or total

Add.

1.	7	4	6	2	7	2	5	3	5	0
	+ 2	+ 4	+ 5	+ 3	+ 1	+ 4	+ 0	+ 9	+ 8	+ 9
	9	8	11	5	8	6	5	12	13	9
2.	3	7	5	6	9	2	3	2	5	1
	+ 6	+ 4	+ 5	+ 4	+ 7	+ 0	+ 3	+ 7	+ 9	+ 4
	9	11	10	10	17	2	6	9	14	5
3.	9	0	1	7	8	7	5	0	6	3
	+ 2	+ 4	+ 6	+ 6	+ 1	+ 5	+ 2	+ 3	+ 9	+ 4
	11	4	7	13	9	12	7	3	15	7
4.	0	1	7	4	9	4	3	6	1	2
	+ 6	+ 5	+ 7	+ 5	+ 0	+ 1	+ 7	+ 2	+ 0	+ 6
	6	6	14	9	9	5	10	8	1	8
5.	4	2	1	8	8	5	6	7	4	6
	+ 8	+ 5	+ 7	+ 8	+ 3	+ 6	+ 0	+ 3	+ 9	+ 8
	12	7	8	16	11	11	6	10	13	14
6.	2	9	3	4	0	3	5	4	1	8
	+ 2	+ 8	+ 5	+ 6	+ 7	+ 2	+ 3	+ 0	+ 3	+ 2
	4	17	8	10	7	5	8	4	4	10
7.	8	5	9	9	1	8	4	8	3	1
	+ 0	+ 7	+ 1	+ 3	+ 2	+ 5	+ 3	+ 6	+ 0	+ 1
	8	12	10	12	3	13	7	14	3	2
8.	4	0	8	6	7	4	5	7	9	0
	+ 7	+ 5	+ 4	+ 3	+ 8	+ 2	+ 1	+ 0	+ 5	+ 2
	11	5	12	9	15	6	6	7	14	2

9.	1 + 8 = 9	9 + 4 = 13	6 + 7 = 13	0 + 1 = 1	1 + 9 = 10	5 + 4 = 9	2 + 8 = 10	9 + 9 = 18	6 + 1 = 7	3 + 8 = 11
10.	6 + 6 = 12	2 + 1 = 3	8 + 9 = 17	2 + 9 = 11	8 + 7 = 15	0 + 8 = 8	9 + 6 = 15	3 + 1 = 4	0 + 0 = 0	7 + 9 = 16
11.	2 + 4 = 8	2 + 0 = 2	7 + 5 = 12	4 + 1 = 5	5 + 6 = 12	3 + 2 = 5	8 + 5 = 13	4 + 2 = 6	5 + 4 = 9	0 + 8 = 8
12.	6 + 5 = 11	5 + 5 = 10	1 + 6 = 7	7 + 7 = 14	1 + 7 = 8	3 + 5 = 8	9 + 1 = 10	8 + 4 = 12	6 + 7 = 13	8 + 9 = 17
13.	5 + 0 = 5	3 + 3 = 6	5 + 2 = 7	3 + 7 = 10	6 + 0 = 6	4 + 3 = 7	5 + 1 = 6	2 + 8 = 10	5 + 3 = 8	9 + 6 = 15
14.	7 + 2 = 9	3 + 6 = 9	9 + 2 = 11	4 + 8 = 12	2 + 2 = 4	8 + 0 = 8	4 + 7 = 11	1 + 8 = 9	6 + 6 = 12	2 + 3 = 5
15.	6 + 4 = 10	7 + 6 = 13	0 + 6 = 6	4 + 5 = 9	8 + 8 = 16	4 + 6 = 10	9 + 3 = 12	6 + 3 = 9	0 + 1 = 1	2 + 9 = 11
16.	0 + 9 = 9	1 + 4 = 5	3 + 4 = 7	2 + 6 = 8	6 + 8 = 15	8 + 2 = 10	1 + 1 = 2	0 + 2 = 2	3 + 8 = 11	7 + 9 = 16
17.	3 + 9 = 12	2 + 7 = 9	0 + 3 = 3	7 + 3 = 10	4 + 0 = 4	8 + 6 = 14	7 + 0 = 7	6 + 2 = 8	9 + 9 = 18	3 + 1 = 4
18.	4 + 4 = 8	7 + 4 = 11	1 + 5 = 6	2 + 5 = 7	9 + 8 = 17	5 + 7 = 12	0 + 5 = 5	9 + 4 = 13	2 + 1 = 3	0 + 4 = 4
19.	7 + 1 = 8	9 + 7 = 16	8 + 1 = 9	9 + 0 = 9	8 + 3 = 10	0 + 7 = 7	1 + 2 = 3	7 + 8 = 15	1 + 9 = 10	8 + 7 = 15
20.	5 + 8 = 13	5 + 9 = 14	6 + 9 = 15	1 + 0 = 1	4 + 9 = 13	1 + 3 = 4	3 + 0 = 3	9 + 5 = 14	6 + 1 = 7	0 + 0 = 0

Adding Larger Numbers

If you know the basic addition facts on pages 15 and 16, you are ready to add larger numbers. Add the column at the right first, and then move to the next column to the left. Continue until you have added each column of figures.

To check an addition problem, you can add the numbers in each column from the bottom.

EXAMPLE

$$\begin{array}{r} 35 \\ +\ 62 \\ \hline 97 \end{array}$$

STEP 1 $5 + 2 = 7$
STEP 2 $3 + 6 = 9$

CHECK

$$\begin{array}{r} 35 \\ +\ 62 \\ \hline 97 \end{array}$$

STEP 1 $2 + 5 = 7$
STEP 2 $6 + 3 = 9$

Add and check.

1.	47 + 42	50 + 38	84 + 15	63 + 24	78 + 10	42 + 56	51 + 22	70 + 19
2.	603 + 285	577 + 321	458 + 201	413 + 564	762 + 125	210 + 648	307 + 652	
3.	4,285 + 3,054	1,756 + 6,213	6,073 + 2,515	4,261 + 3,428	2,352 + 6,043	7,084 + 2,713		
4.	408 + 241	69,043 + 20,516	27 + 40	523 + 271	861,042 + 121,446	56 + 22		
5.	7,244 + 2,351	8,563 + 1,234	5,042 + 2,635	7,136 + 1,042	3,574 + 3,103	5,546 + 2,342		
6.	8,027 + 1,932	3,329 + 4,060	6,203 + 2,351	4,815 + 2,114	8,047 + 1,832	1,966 + 4,032		
7.	6,001 + 2,573	2,347 + 2,412	7,580 + 1,316	4,116 + 4,572	7,038 + 2,521	2,413 + 4,334		
8.	24,207 + 15,072	54,156 + 30,422	32,854 + 43,104	25,401 + 22,367	69,821 + 20,104			

Adding and Carrying

When the sum of the numbers in a column has a 2-digit answer (such as 13 in Step 1 in the example below), write the digit on the right under the column you just added and carry the left digit to the next column.

Writing a digit in the next column is sometimes called **regrouping** or **renaming.** It is also called **carrying.** The idea is to add units with units, tens with tens, hundreds with hundreds, and so on.

EXAMPLE

$$\begin{array}{r} {\scriptstyle 1\,1} \\ 546 \\ +\ 297 \\ \hline 843 \end{array}$$

STEP 1 $6 + 7 = 13$. Write 3 in the units column and 1 above the tens column.

STEP 2 $1 + 4 + 9 = 14$. Write 4 in the tens column and 1 above the hundreds column.

STEP 3 $1 + 5 + 2 = 8$

To check an addition problem, you can add from the bottom as you learned on page 17. You can also try the following method.

$$\begin{array}{r} 546 \\ +\ 297 \\ \hline 843 \end{array}$$

		Write down
STEP 1	Add the first column on the right in your head and write the *entire* total. $6 + 7 = 13$	13
STEP 2	Add the next column to the left and write that *entire* total under the first total and one column to the left. $4 + 9 = 13$	$\begin{array}{r} 13\ \\ 13\ \ \end{array}$
STEP 3	Add the next column to the left and write that *entire* total under the second total and one more column to the left. $5 + 2 = 7$	$\begin{array}{l} \ \ 13 \\ \ 13 \\ 7 \end{array}$
STEP 4	Add the three totals.	843

This method can be used with any number of figures. Always remember to write the total of each column under the previous column *and* one column to the left.

Add and check. The first problem has been done for you.

	a	b	c	d	e	f
1.	46 + 47 = 93 (check: 13, 8, 93)	57 + 83	67 + 54	43 + 29	82 + 68	29 + 41
2.	38 + 93	63 + 29	22 + 78	54 + 99	48 + 53	73 + 89
3.	683 + 417	257 + 683	594 + 417	267 + 936	607 + 406	219 + 389
4.	384 + 429	968 + 518	314 + 98	283 + 56	607 + 95	966 + 35
5.	748 + 87	825 + 59	346 + 64	293 + 53	53 + 469	64 + 377
6.	68 + 580	72 + 469	93 + 708	87 + 506	41 + 289	57 + 418
7.	1 + 30 + 9	5 + 47 + 8	2 + 81 + 5	3 + 25 + 6	2 + 56 + 5	9 + 37 + 8
8.	6 + 52 + 9	5 + 83 + 4	66 + 81 + 56	21 + 49 + 43	89 + 67 + 50	92 + 10 + 39
9.	87 + 19 + 72	60 + 54 + 38	31 + 28 + 44	79 + 96 + 83	20 + 74 + 627	95 + 80 + 417

10.	70	51	42	78	63	84
	57	91	93	12	85	43
	+ 202	+ 329	+ 516	+ 490	+ 854	+ 977

11.	365	236	413	146	243	880
	520	153	648	73	10	77
	+ 124	+ 875	+ 381	+ 718	+ 256	+ 523

12.	91	15	23	646	930	347
	31	75	22	60	15	38
	78	76	87	52	37	67
	+ 66	+ 31	+ 99	+ 994	+ 376	+ 421

13.	685	597	263	760	461	298
	691	283	161	218	919	709
	274	406	247	322	653	395
	+ 394	+ 938	+ 459	+ 938	+ 597	+ 471

14.	57	43	86	94	40	27
	35	51	65	21	83	40
	12	74	84	39	26	67
	83	32	46	45	29	28
	+ 58	+ 36	+ 97	+ 80	+ 33	+ 86

15.	518	114	806	638	441
	782	726	992	793	348
	764	953	528	136	635
	207	199	666	483	914
	+ 843	+ 727	+ 894	+ 397	+ 679

16.	8,779	4,855	7,630	4,596
	2,286	2,849	4,108	8,892
	+ 5,269	+ 1,754	+ 7,068	+ 4,625

17.	3,948	6,787	4,108	9,081
	7,758	3,316	7,915	8,752
	6,799	4,213	3,736	2,978
	+ 2,437	+ 5,449	+ 2,615	+ 7,093

18.	520	714	863	933
	2,186	7,465	6,097	9,487
	706	302	141	761
	5,807	4,374	8,262	3,309
	+ 935	+ 491	+ 516	+ 743

Adding Numbers Written Horizontally

When the numbers you want to add are not in vertical columns, rewrite them so that the units are under the units, the tens are under the tens, and so on. Always line up the units column *first*.

EXAMPLE $25 + 6 + 423 =$

REWRITE AS

$$\begin{array}{r} 25 \\ 6 \\ +\ 423 \\ \hline 454 \end{array}$$

To add money, be sure to add dollars in the dollars columns and cents in the cents columns. For example, to find the sum of \$4, \$2.95, and 36¢, set the numbers under each other in this way:

$$\begin{array}{r} \$4.00 \\ 2.95 \\ +\ 0.36 \\ \hline \$7.31 \end{array}$$

Add and check.

1. $93 + 55 + 34 =$ $\qquad$ $23 + 18 + 96 =$ $\qquad$ \$4 + \$5.19 =

2. $607 + 12 + 344 =$ $\qquad$ $58 + 752 + 29 =$ $\qquad$ \$1.28 + \$8.99 + \$6 =

3. $53 + 618 + 9 + 47 =$ $\qquad$ $7 + 24 + 806 + 63 =$ $\qquad$ 14¢ + \$9 + \$7.60 =

4. $982 + 3{,}507 + 73 + 184 =$ $\qquad$ $26 + 745 + 66 + 4{,}329 =$ $\qquad$ \$17.50 + \$6 + \$4.83 =

5. $8{,}196 + 75{,}883 + 29 + 334 =$ $\qquad$ $232 + 80{,}465 + 19 + 1{,}591 =$ $\qquad$ \$0.16 + \$4.55 + \$3 =

6. $8{,}779 + 2{,}286 + 5{,}269 =$ $\qquad$ $4{,}855 + 2{,}849 + 1{,}754 =$ $\qquad$ \$2.85 + 36¢ + \$10 =

Addition Shortcuts

Add each problem in your head.

1. $42 + 8 =$ $\quad$ $23 + 7 =$ $\quad$ $3 + 87 =$

2. $76 + 4 =$ $\quad$ $1 + 59 =$ $\quad$ $2 + 68 =$

3. $9 + 31 =$ $\quad$ $6 + 14 =$ $\quad$ $65 + 5 =$

The answer to each of the previous problems ends with zero. Two numbers whose sum ends in zero are sometimes called **compatible pairs.**

Look at the example below. You start to add $13 + 9 = 22$ but notice that $9 + 11 = 20$. Add these numbers first. Remember that *the numbers in an addition problem may be added in any order.* Look for a compatible pair, and add the pair first.

EXAMPLE

$13 + 9 + 11 =$ The numbers 9 and 11 are a compatible pair. $9 + 11 = 20$

$13 + 20 = 33$ Rewrite the problem as the sum of $13 + 20$.

Find a compatible pair in each problem. Then rewrite the problem, and add.

4. $62 + 8 + 5 =$ $\quad$ $18 + 45 + 5 =$

5. $19 + 13 + 7 =$ $\quad$ $6 + 94 + 13 =$

6. $36 + 12 + 4 =$ $\quad$ $52 + 8 + 21 =$

7. $9 + 14 + 51 =$ $\quad$ $19 + 27 + 3 =$

Find two compatible pairs in each problem. Then rewrite the problem, and add.

8. $28 + 17 + 2 + 3 + 9 =$ $\quad$ $12 + 9 + 6 + 21 + 8 =$

9. $6 + 35 + 11 + 14 + 5 =$ $\quad$ $4 + 35 + 13 + 26 + 5 =$

10. $15 + 19 + 1 + 27 + 3 =$ $\quad$ $21 + 14 + 42 + 8 + 6 =$

11. $43 + 7 + 8 + 62 + 5 =$ $\quad$ $17 + 11 + 5 + 9 + 3 =$

Rounding and Estimating

Review the rules for rounding numbers on page 11. When the digit to the right of the number you are rounding to is greater than or equal to 5, you must add 1 to the underlined digit. If the underlined digit is a 9, the digit to the left of 9 changes. Look at this example carefully.

EXAMPLE Round 496 to the nearest ten. 4<u>9</u>6 → 500

STEP 1 Underline 9, the digit in the tens place.

STEP 2 The digit to the right of 9 is 6. Since 6 is greater than 5, add 1 to 9. $9 + 1 = 10$
Write 0 in the tens column, and add 1 to the hundreds column.
$1 + 4 = 5$

STEP 3 Change the digit in the units place to 0.

Notice, in the last example, that when you round 496 to the nearest ten, you get 500. When you round 496 to the nearest hundred, you also get 500.

Round each number to the nearest *ten.*

1. 296	98	3,095	649	4,583	12,487

Round each number to the nearest *hundred.*

2. 3,972	24,968	8,391	7,083	42,952	6,546

Round each number to the nearest *thousand.*

3. 49,806	75,928	219,655	8,962	69,506	2,384

Estimation means finding a reasonable answer. An estimate is not exact, but it is close. Rounding the numbers in a problem is a good way to estimate an answer.

Round each number to the nearest *hundred.* Then add the rounded numbers. The symbol ≈ means "approximately equal to."

4. $3{,}948 + 758 + 6{,}799 + 437 \approx$ $9{,}081 + 752 + 2{,}978 + 93 \approx$

5. $787 + 3{,}316 + 213 + 5{,}449 \approx$ $520 + 2{,}186 + 706 + 5{,}807 \approx$

6. $108 + 7{,}915 + 3{,}736 + 615 \approx$ $714 + 7{,}465 + 302 + 4{,}974 \approx$

Round each number to the nearest *thousand.* Then add the rounded numbers.

7. 76,493 + 6,590 + 27,286 ≈ 19,863 + 6,097 + 28,141 ≈

8. 4,538 + 64,908 + 70,435 ≈ 7,933 + 9,487 + 16,761 ≈

9. 27,881 + 92,855 + 3,064 ≈ 41,935 + 2,491 + 18,516 ≈

Another way to estimate is to use **front-end rounding.** To get a quick estimate of an answer, round each number in a problem to the *left-most* place. Then add the rounded numbers.

EXAMPLE **Use front-end rounding to estimate the answer to the problem 52 + 873 + 2,184.**

52 + 873 + 2,184 ≈ 50 + 900 + 2,000 =	**STEP 1**	Round 52 to the nearest ten, 873 to the nearest hundred, and 2,184 to the nearest thousand.
2,950	**STEP 2**	Add the rounded numbers. The exact answer is 3,109.

For problems 10 to 15, use front-end rounding to estimate each answer.

10. 18 + 2,936 + 520 ≈ 4,297 + 58 + 316 ≈

11. 114 + 3,726 + 83 ≈ 199 + 8,066 + 12 ≈

12. 6,806 + 992 + 94 ≈ 54 + 8,705 + 684 ≈

13. 596,868 + 43,117 + 6,462 ≈ 95,048 + 1,289 + 472,016 ≈

14. 88,273 + 2,897 + 659,084 ≈ 1,095 + 19,708 + 182,550 ≈

15. 6,287 + 203,900 + 4,951 ≈ 4,680 + 790,523 + 5,407 ≈

Applying Your Addition Skills

The problems in this exercise require that you apply your addition skills to practical problems. Pay close attention to the language in each problem. Watch for words such as **sum, total, combined,** and **altogether.** These words suggest addition.

Estimating is a good way to decide whether an answer is reasonable. After many problems in this exercise, you will be asked to round and estimate the answer to the problems you solved.

Solve each problem, and write the correct label such as $, miles, or pounds next to each answer.

1. Alfonso drives a truck four days a week. Last week he drove 308 miles on Monday, 397 miles on Tuesday, 389 miles on Wednesday, and 286 miles on Thursday. How many miles altogether did Alfonso drive last week?

2. To estimate the distance Alfonso drove his truck last week, round each distance to the nearest *hundred* miles. Use the rounded numbers to find the approximate distance that he drove.

3. Every summer Laura and Tim run an ice cream shop in the town where they have a lakeside cottage. Last summer they had sales of $4,186 in June, $6,709 in July, $10,682 in August, and $1,948 in September. Find their total sales for last summer.

4. Estimate Laura and Tim's total sales for last summer by first rounding each month's sales to the nearest *thousand.*

5. A group of friends started a food-delivery business. Mr. Reilly invested $17,680, Mr. Aladro invested $14,275, and Mr. Chan invested $18,923. Find the combined amount of their investments.

6. For the friends in the last problem, round each investment to the nearest *thousand* dollars. Then use the rounded numbers to estimate their combined investments.

7. Attendance at a three-day basketball tournament was the following: Thursday, 2,640; Friday, 2,873; and Saturday, 3,084. What was the total attendance at the tournament?

8. Estimate the total attendance at the basketball tournament by first rounding each night's attendance to the nearest *hundred.*

9. Michelle works at the check-out counter of the Lake Pharmacy. When the register broke down, she had to add the day's receipts by hand. Purchases that day came to a total of $650.38 for prescription drugs, $578.12 for over-the-counter medications, $386.52 for household items, and $498.84 for cosmetics. What was the sum of the receipts for the day?

10. Round each category of sales in the last problem to the nearest *ten* dollars. Then estimate the day's total with the rounded numbers.

11. Pete loaded six boxes onto a freight elevator. One box weighed 165 pounds; the second weighed 218 pounds; the third, 186 pounds; the fourth, 197 pounds, the fifth, 175 pounds; and the sixth, 229 pounds. Find the combined weight of the boxes.

12. The freight elevator in the last problem can carry 1,250 pounds. Can the elevator safely carry all the boxes?

13. Pete, in problem 11, weighs 210 pounds. Can he safely ride in the freight elevator with the load of boxes?

14. Yiu decided to go off his diet one afternoon, so he ate his favorite foods. For a snack he had a ginger ale (75 calories) and a bag of pretzel sticks (190 calories). Later he ate a cheese pizza (555 calories) and drank a cola (95 calories). Find the total number of calories Yiu consumed that afternoon.

15. Estimate the number of calories that Yiu consumed by first rounding each calorie amount to the nearest *ten.*

The next problems include large numbers that need to be rounded. Large numbers are often less accurate than they seem to be. For example, the exact population of a big city can be accurate for only a short period of time. Large numbers are more useful when they end in zeros.

Use the following information to answer questions 16 to 18.

The following numbers represent active military personnel in the United States in a recent year.

Army	493,825
Navy/Marines	523,658
Air Force	368,825

16. What was the combined number of active military personnel that year?

17. Round the number of military personnel in each branch to the nearest *thousand* to estimate the total number serving that year.

18. Round the number of military personnel in each branch to the nearest *ten thousand* to estimate the total number serving that year.

Use the following information to answer questions 19 and 20.

The area of each of the five Great Lakes is as follows: Ontario, 7,540 square miles; Erie, 9,940 square miles; Michigan, 22,400 square miles; Huron, 23,010 square miles; and Superior, 31,820 square miles.

19. What is the combined area of the five Great Lakes?

20. To estimate the combined area of the Great Lakes, first round the area of each lake to the nearest *thousand* square miles.

Use the following information to answer questions 21 and 22.

According to the Census Bureau, the population of each of the five boroughs of New York City in a recent year was as follows:

Borough	Population
Bronx	1,332,650
Brooklyn	2,465,326
Manhattan	1,537,195
Queens	2,229,370
Staten Island	443,728

21. What was the total population of New York City that year?

22. Round the population of each borough to the nearest *hundred thousand* to estimate the total population of New York City that year.

The next problems do not have any of the common key words that suggest addition. Read each problem carefully to get a sense of how the problem asks you to add.

23. When David sold his car to his younger sister, he got $2,750 for it. This was $1,245 less than he paid for the car 18 months earlier. How much did David pay for the car?

24. To estimate the amount David paid for his car, first round the numbers to the nearest *hundred* dollars.

25. The distance from Cincinnati to Columbus is 108 miles. The distance from Columbus to Cleveland is 133 miles. What is the distance from Cincinnati to Cleveland by way of Columbus?

26. Anna's great-grandfather built his house in 1897. The house was demolished to make way for a shopping center 105 years later. In what year was the house demolished?

Addition Review

This review covers the first two chapters of this book. When you finish, check your answers at the back of the book.

1. What is the value of the digit 2 in the number 342,508?

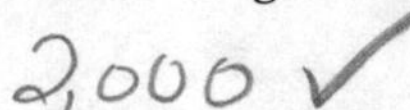

2. What is 14,763 rounded to the nearest *hundred?*

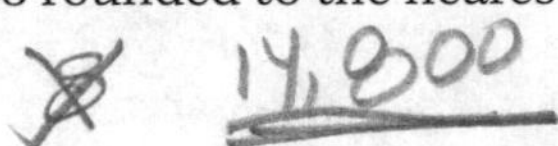

3. What is $481,297 rounded to the nearest *ten thousand?*

4. $5{,}324 + 2{,}145$

5. $23{,}864 + 51{,}133$

6. $348{,}065 + 251{,}822$

7. $56 + 87 + 25$

8. $63 + 54 + 97 + 81$

9. $32 + 40 + 76 + 51 + 49$

10. $816 + 53 + 228 + 46$

11. $53 + 226 + 45 + 513$

12. $619 + 58 + 417 + 96$

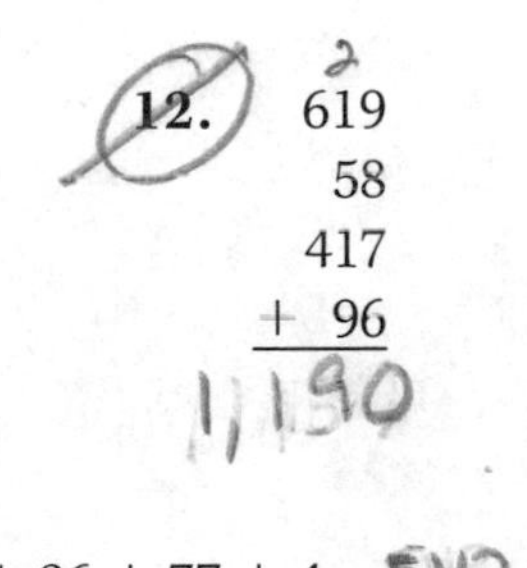

13. $63 + 214 + 55 =$

14. $425 + 36 + 77 + 4 =$

15. $817 + 1{,}329 + 638 =$

16. $12 + 3{,}407 + 8 + 496 =$

17. $796{,}041 + 532{,}408 =$

18. $27 + 48 + 3 + 901 + 2 =$

19. Round each number to the nearest *hundred* and add.

$3{,}972 + 1{,}258 + 3{,}417 \approx$

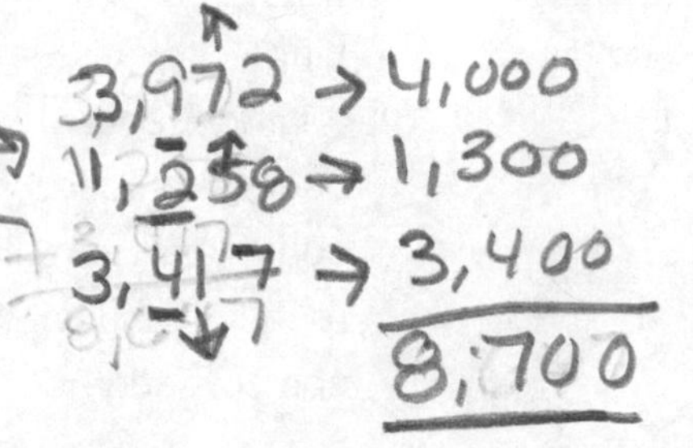

20. Round each number to the nearest *thousand* and add.

69,836 + 71,254 + 33,927 ≈

21. The population of Baltimore is 637,418. Round the number to the nearest *ten thousand.*

22. In a month Tiva paid $575.00 for rent, $38.94 for gas and electricity, $65.47 for telephone, and $110.00 for her car payment. Find the total of these expenses.

23. Uncle Bob's car dealership sold 483 cars the first year it was open. The second year the dealership sold 238 more than the first year. Find the total sales for the second year.

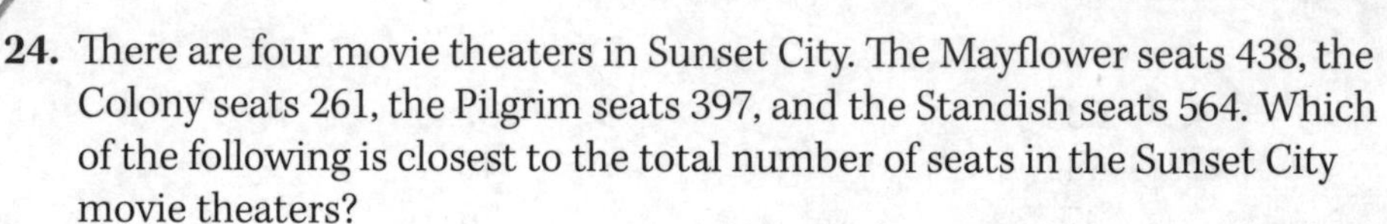

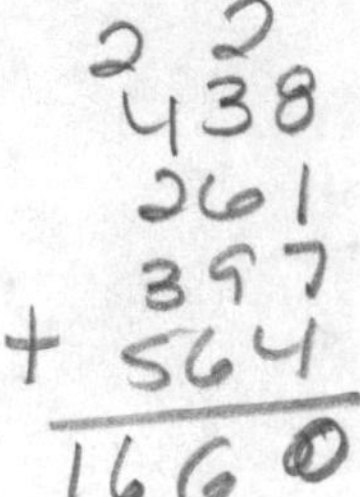

24. There are four movie theaters in Sunset City. The Mayflower seats 438, the Colony seats 261, the Pilgrim seats 397, and the Standish seats 564. Which of the following is closest to the total number of seats in the Sunset City movie theaters?

a. 1,300 seats
b. 1,500 seats
c. 1,700 seats
d. 1,900 seats

ADDITION REVIEW CHART

If you missed more than one problem on any group below, review the practice pages for those problems. Then redo the problems you got wrong before going on to the Subtraction Skills Inventory. If you had a passing score, redo any problem you missed and begin the Subtraction Skills Inventory on page 31.

Problem Numbers	Skill Area	Practice Pages
1, 2, 3	place value	6–12
4, 5, 6	addition facts	15–17
7, 8, 9, 10, 11, 12	adding and carrying	18–20
13, 14, 15, 16, 17, 18	adding horizontally	21
19, 20, 21	rounding and estimating	23–24
22, 23, 24	applying addition	25–28

SUBTRACTION

Subtraction Skills Inventory

Do all the problems that you can. There is no time limit. Work carefully and check your answers, but do not use outside help. Correct answers are listed by page number at the back of the book.

1. $74 - 30$

2. $58 - 27$

3. $542 - 241$

4. $759 - 207$

5. $47 - 29$

6. $764 - 19$

7. $9{,}183 - 2{,}097$

8. $4{,}167 - 2{,}758$

9. $402 - 358$

10. $500 - 276$

11. $3{,}008 - 1{,}956$

12. $10{,}000 - 7{,}049$

13. 528 − 39 =

14. 425 − 276 =

15. 23,532 − 6,548 =

16. 70,060 − 4,118 =

17. 110 − 64 =

18. 73 − 25 =

19. Round each number to the nearest *hundred* and subtract.

7,498 − 2,379 ≈

20. Round each number to the nearest *thousand* and subtract.

59,614 − 29,058 ≈

21. Round each number to the nearest *ten thousand* and subtract.

485,311 − 296,347 ≈

22. Joe's gross salary is $619.42 per month. His employer deducts $156.75 from Joe's paycheck for taxes and benefits. What is Joe's take-home pay?

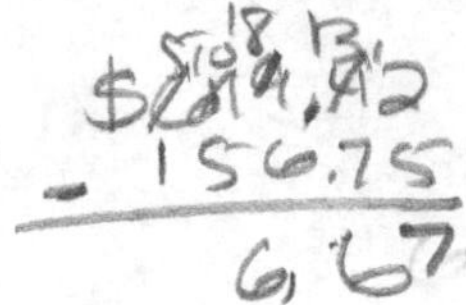

23. Mr. Greenwald bought a chair for $86.95. How much change did he receive if he paid with a $100 bill?

24. Before the Walek family went on vacation, their odometer (mileage dial) had a reading of 34,925 miles. When they returned, it read 36,059 miles. How far did they drive on their vacation?

25. Clarktown has 3,595 registered voters. At the last election, 2,459 people voted in Clarktown. How many of the registered voters did not vote?

SUBTRACTION SKILLS INVENTORY CHART

If you missed more than one problem on any group below, work through the practice pages for that group. Then redo the problems you got wrong on the Subtraction Skills Inventory. If you had a passing score on all seven groups of problems, redo any problem you missed and begin the Multiplication Skills Inventory on page 54.

Problem Numbers	Skill Area	Practice Pages
1, 2, 3, 4	subtraction facts	33–35
5, 6, 7, 8	subtracting and borrowing	36–38
9, 10, 11, 12	subtracting from zeros	40–42
13, 14, 15, 16	subtracting horizontally	39, 42
17, 18	subtraction shortcuts	43
19, 20, 21	rounding and estimating	44
22, 23, 24, 25	applying subtraction	45–50

Basic Subtraction Facts

This page and the following page will help you learn the basic subtraction facts. The subtraction facts are basic building blocks for further study of mathematics. You must *memorize* these facts. Check your answers to this exercise. Practice the facts you got wrong. Then try these problems again until you can get all the answers quickly and accurately.

Parts of a Subtraction Problem

$$\begin{array}{rl} 6 & \leftarrow \text{minuend} \\ -2 & \leftarrow \text{subtrahend} \\ \hline 4 & \leftarrow \text{difference} \end{array}$$

1.	$\begin{array}{r} 5 \\ -3 \\ \hline 2 \end{array}$	$\begin{array}{r} 7 \\ -6 \\ \hline 1 \end{array}$	$\begin{array}{r} 8 \\ -3 \\ \hline 5 \end{array}$	$\begin{array}{r} 13 \\ -5 \\ \hline 8 \end{array}$	$\begin{array}{r} 11 \\ -3 \\ \hline 8 \end{array}$	$\begin{array}{r} 7 \\ -2 \\ \hline 5 \end{array}$	$\begin{array}{r} 8 \\ -0 \\ \hline 8 \end{array}$	$\begin{array}{r} 11 \\ -9 \\ \hline 2 \end{array}$	$\begin{array}{r} 10 \\ -7 \\ \hline 3 \end{array}$	$\begin{array}{r} 9 \\ -9 \\ \hline 0 \end{array}$
2.	$\begin{array}{r} 7 \\ -1 \\ \hline 6 \end{array}$	$\begin{array}{r} 5 \\ -2 \\ \hline 3 \end{array}$	$\begin{array}{r} 13 \\ -7 \\ \hline 6 \end{array}$	$\begin{array}{r} 14 \\ -6 \\ \hline 8 \end{array}$	$\begin{array}{r} 3 \\ -2 \\ \hline 1 \end{array}$	$\begin{array}{r} 4 \\ -3 \\ \hline 1 \end{array}$	$\begin{array}{r} 15 \\ -6 \\ \hline 9 \end{array}$	$\begin{array}{r} 10 \\ -9 \\ \hline 1 \end{array}$	$\begin{array}{r} 9 \\ -3 \\ \hline 6 \end{array}$	$\begin{array}{r} 2 \\ -1 \\ \hline 1 \end{array}$
3.	$\begin{array}{r} 6 \\ -4 \\ \hline 2 \end{array}$	$\begin{array}{r} 5 \\ -0 \\ \hline 5 \end{array}$	$\begin{array}{r} 10 \\ -2 \\ \hline 8 \end{array}$	$\begin{array}{r} 13 \\ -4 \\ \hline 9 \end{array}$	$\begin{array}{r} 8 \\ -5 \\ \hline 3 \end{array}$	$\begin{array}{r} 17 \\ -8 \\ \hline 9 \end{array}$	$\begin{array}{r} 7 \\ -4 \\ \hline 3 \end{array}$	$\begin{array}{r} 3 \\ -3 \\ \hline 0 \end{array}$	$\begin{array}{r} 15 \\ -9 \\ \hline 6 \end{array}$	$\begin{array}{r} 11 \\ -8 \\ \hline 3 \end{array}$
4.	$\begin{array}{r} 1 \\ -1 \\ \hline 0 \end{array}$	$\begin{array}{r} 14 \\ -5 \\ \hline 9 \end{array}$	$\begin{array}{r} 6 \\ -3 \\ \hline 3 \end{array}$	$\begin{array}{r} 8 \\ -7 \\ \hline 1 \end{array}$	$\begin{array}{r} 9 \\ -6 \\ \hline 3 \end{array}$	$\begin{array}{r} 15 \\ -8 \\ \hline 7 \end{array}$	$\begin{array}{r} 11 \\ -6 \\ \hline 5 \end{array}$	$\begin{array}{r} 10 \\ -6 \\ \hline 4 \end{array}$	$\begin{array}{r} 4 \\ -4 \\ \hline 0 \end{array}$	$\begin{array}{r} 12 \\ -7 \\ \hline 5 \end{array}$
5.	$\begin{array}{r} 9 \\ -2 \\ \hline 7 \end{array}$	$\begin{array}{r} 3 \\ -0 \\ \hline 3 \end{array}$	$\begin{array}{r} 12 \\ -5 \\ \hline 7 \end{array}$	$\begin{array}{r} 8 \\ -8 \\ \hline 0 \end{array}$	$\begin{array}{r} 9 \\ -4 \\ \hline 5 \end{array}$	$\begin{array}{r} 10 \\ -3 \\ \hline 7 \end{array}$	$\begin{array}{r} 13 \\ -6 \\ \hline 7 \end{array}$	$\begin{array}{r} 7 \\ -7 \\ \hline 0 \end{array}$	$\begin{array}{r} 12 \\ -8 \\ \hline 4 \end{array}$	$\begin{array}{r} 8 \\ -4 \\ \hline 2 \end{array}$
6.	$\begin{array}{r} 10 \\ -1 \\ \hline 9 \end{array}$	$\begin{array}{r} 14 \\ -9 \\ \hline 5 \end{array}$	$\begin{array}{r} 12 \\ -6 \\ \hline 6 \end{array}$	$\begin{array}{r} 6 \\ -0 \\ \hline 6 \end{array}$	$\begin{array}{r} 7 \\ -5 \\ \hline 2 \end{array}$	$\begin{array}{r} 11 \\ -7 \\ \hline 4 \end{array}$	$\begin{array}{r} 9 \\ -5 \\ \hline 4 \end{array}$	$\begin{array}{r} 10 \\ -4 \\ \hline 6 \end{array}$	$\begin{array}{r} 1 \\ -0 \\ \hline 1 \end{array}$	$\begin{array}{r} 8 \\ -1 \\ \hline 7 \end{array}$
7.	$\begin{array}{r} 16 \\ -7 \\ \hline 9 \end{array}$	$\begin{array}{r} 8 \\ -6 \\ \hline 2 \end{array}$	$\begin{array}{r} 10 \\ -5 \\ \hline 5 \end{array}$	$\begin{array}{r} 5 \\ -4 \\ \hline 1 \end{array}$	$\begin{array}{r} 6 \\ -2 \\ \hline 4 \end{array}$	$\begin{array}{r} 5 \\ -1 \\ \hline 4 \end{array}$	$\begin{array}{r} 7 \\ -3 \\ \hline 4 \end{array}$	$\begin{array}{r} 14 \\ -8 \\ \hline 6 \end{array}$	$\begin{array}{r} 18 \\ -9 \\ \hline 9 \end{array}$	$\begin{array}{r} 11 \\ -4 \\ \hline 7 \end{array}$
8.	$\begin{array}{r} 4 \\ -0 \\ \hline 4 \end{array}$	$\begin{array}{r} 8 \\ -2 \\ \hline 6 \end{array}$	$\begin{array}{r} 13 \\ -8 \\ \hline 5 \end{array}$	$\begin{array}{r} 9 \\ -1 \\ \hline 8 \end{array}$	$\begin{array}{r} 6 \\ -6 \\ \hline 0 \end{array}$	$\begin{array}{r} 12 \\ -3 \\ \hline 9 \end{array}$	$\begin{array}{r} 3 \\ -1 \\ \hline 2 \end{array}$	$\begin{array}{r} 15 \\ -7 \\ \hline 8 \end{array}$	$\begin{array}{r} 11 \\ -5 \\ \hline 6 \end{array}$	$\begin{array}{r} 4 \\ -1 \\ \hline 3 \end{array}$

9.	9 − 7	2 − 2	14 − 7	16 − 9	7 − 0	11 − 2	5 − 5	13 − 9	2 − 0	12 − 4
	2	0	7	7	7	9	0	4	0	8
10.	6 − 5	17 − 9	9 − 8	6 − 1	16 − 8	12 − 9	9 − 0	4 − 2	10 − 8	13 − 6
	1	8	1	5	8	3	9	2	2	7
11.	8 − 3	7 − 2	10 − 7	5 − 2	3 − 2	10 − 9	6 − 4	13 − 4	7 − 4	11 − 8
	5	5	3	3	1	1	2	9	3	3
12.	6 − 3	15 − 8	4 − 4	3 − 0	9 − 4	7 − 7	10 − 1	6 − 0	9 − 5	8 − 1
	3	7	0	3	5	0	9	6	4	7
13.	10 − 5	5 − 1	18 − 9	8 − 2	6 − 6	15 − 7	9 − 7	16 − 9	5 − 5	12 − 4
	5	4	9	6	0	8	2	7	0	8
14.	9 − 8	12 − 9	10 − 8	7 − 6	11 − 3	11 − 9	7 − 1	14 − 6	15 − 6	2 − 1
	1	3	2	1	8	2	6	8	9	1
15.	10 − 2	17 − 8	15 − 9	14 − 5	9 − 6	10 − 6	9 − 2	8 − 8	13 − 6	8 − 4
	8	9	6	9	3	4	1	0	8	4
16.	12 − 6	11 − 7	1 − 0	8 − 6	6 − 2	14 − 8	4 − 0	9 − 1	3 − 1	4 − 1
	6	4	1	2	4	6	4	8	2	3
17.	14 − 7	11 − 2	2 − 0	17 − 9	16 − 8	4 − 2	5 − 3	13 − 5	8 − 0	9 − 9
	7	9	2	8	8	2	2	8	8	0
18.	13 − 7	4 − 3	9 − 3	5 − 0	8 − 5	3 − 3	1 − 1	8 − 7	11 − 6	12 − 7
	6	1	6	5	3	0	0	1	5	5
19.	12 − 5	10 − 3	12 − 8	14 − 9	7 − 5	10 − 4	16 − 7	5 − 4	7 − 3	11 − 4
	7	7	4	5	2	6	9	1	4	7
20.	13 − 8	12 − 3	11 − 5	2 − 2	7 − 0	13 − 9	6 − 5	6 − 1	9 − 0	13 − 6
	5	9	6	0	7	4	1	5	9	7

Subtracting Larger Numbers

If you know the basic subtraction facts on pages 33 and 34, you are ready to subtract larger numbers. Subtract the column at the right first, and then move to the next column to the left. Continue until you have subtracted each column of figures.

To check a subtraction problem, add the answer to the bottom number of the original problem. The sum should be the top number of the original problem.

EXAMPLE

$$\begin{array}{r} 46 \\ -\ 25 \\ \hline 21 \end{array}$$

STEP 1 $6 - 5 = 1$
STEP 2 $4 - 2 = 2$

CHECK

$$\begin{array}{r} 46 \\ -\ 25 \\ \hline 21 \\ \hline 46\ ✓ \end{array}$$

STEP 1 $5 + 1 = 6$
STEP 2 $2 + 2 = 4$

Subtract and check.

1. 34 − 22   69 − 35   38 − 31   52 − 32   75 − 51   81 − 40   62 − 11   79 − 58

2. 60 − 40   35 − 21   77 − 73   74 − 40   58 − 32   41 − 21   37 − 15   62 − 21

3. 249 − 134   553 − 243   916 − 503   692 − 491   485 − 324   727 − 512   609 − 203

4. 887 − 352   543 − 243   781 − 300   942 − 611   889 − 278   468 − 432   982 − 581

5. 8,559 − 4,208   7,388 − 6,234   9,161 − 5,031   9,798 − 9,624   9,897 − 3,572   7,258 − 6,125

6. 7,245 − 7,123   4,278 − 3,072   2,657 − 1,452   6,867 − 3,260   7,886 − 4,253   7,980 − 3,210

7. 86,947 − 32,825   56,739 − 54,231   92,465 − 22,143   75,826 − 23,521   59,428 − 41,314

8. 787,445 − 257,312   255,694 − 241,454   860,956 − 360,224   568,798 − 325,416   874,395 − 211,243

Subtracting and Borrowing

When a digit in the subtrahend (the bottom number) is too large to subtract from the digit above it, you have to **borrow** from the next column to the left in the minuend (the top number). Borrowing is sometimes called **regrouping** or **renaming.**

EXAMPLE

$$\begin{array}{r} {}^{8}\not{9}{}^{1}3 \\ -16 \\ \hline 77 \end{array}$$

STEP 1 Since you cannot subtract 6 from 3, borrow 1 ten from the tens column. (10 + 3 = 13)
Put a small 1 next to the 3 in the top number to show that it is now 13.

STEP 2 Cross out the 9 in the tens column and make it an 8 to show that you have borrowed 1 ten.

STEP 3 Subtract the units 13 − 6 = 7.

STEP 4 Subtract the tens 8 − 1 = 7.

CHECK

$$\begin{array}{r} 93 \\ -16 \\ \hline 77 \\ \hline 93 \end{array} \; \checkmark$$

(16 and 77 are bracketed together with a "+".)

STEP 5 Check 16 + 77 = 93.

Subtract and check.

1.	82 − 9	27 − 8	21 − 7	95 − 8	63 − 5	91 − 6	62 − 9	33 − 4
2.	83 − 25	95 − 39	50 − 28	68 − 49	97 − 58	52 − 27	45 − 16	64 − 26
3.	251 − 38	742 − 27	927 − 18	381 − 53	793 − 86	640 − 27	365 − 48
4.	371 − 269	684 − 547	143 − 128	467 − 349	355 − 237	653 − 409	266 − 118
5.	458 − 349	691 − 227	854 − 536	794 − 478	652 − 239	498 − 379	385 − 166

Borrowing More Than Once

Sometimes you have to borrow more than once in the same problem. Many of the problems on this page and the following two pages will give you practice with this kind of subtraction problem.

EXAMPLE

```
   1¹1              3¹2 ¹1 ¹3
   2̸2̸¹3             4̸3̸,2̸4̸¹1
 − 178            − 28,985
    45              14,256
   223 ✓            43,241 ✓
```

Subtract and check.

1.	968 − 74	141 − 80	572 − 91	543 − 62	519 − 53	328 − 86
2.	525 − 247	287 − 198	216 − 158	466 − 388	373 − 199	983 − 585
3.	3,633 − 794	4,125 − 636	2,961 − 972	5,648 − 649	6,345 − 258	1,827 − 918
4.	7,792 − 4,829	8,235 − 3,516	2,047 − 1,839	5,167 − 2,758	4,660 − 1,857	7,133 − 2,924
5.	2,328 − 1,489	5,181 − 2,397	8,760 − 5,764	7,395 − 2,498	7,227 − 4,285	5,684 − 2,993
6.	4,871 − 2,984	6,524 − 3,576	2,775 − 1,799	4,726 − 3,946	8,673 − 1,888	2,865 − 1,977

7.	42,356 − 9,485	38,182 − 8,491	87,421 − 9,530	78,253 − 9,682	34,818 − 5,847
8.	92,309 − 8,958	69,762 − 4,974	86,583 − 5,999	47,757 − 6,858	81,912 − 1,943
9.	93,532 − 9,544	83,124 − 4,876	41,591 − 5,694	35,876 − 8,977	66,180 − 7,293
10.	74,925 − 23,936	53,286 − 43,689	42,755 − 22,697	54,278 − 32,489	45,188 − 21,279
11.	22,483 − 19,584	62,537 − 56,849	55,130 − 43,273	25,614 − 16,726	28,786 − 19,797
12.	678,054 − 97,347	327,583 − 58,946	227,150 − 82,366	732,122 − 51,475	934,662 − 66,539
13.	923,644 − 849,357 74,287 + 849,357 923,644	482,637 − 297,148 185,489 + 297,148 482,637	822,472 − 631,584 190,888 + 631,584 482,637	673,612 − 297,641 185,489 + 297,148 673,112	834,124 − 535,217 298,907 535,217 834,124
14.	543,950 − 418,513 125,437 + 418,513 543,950	437,032 − 214,877 222,155 + 214,877 437,032	144,480 − 142,895 1,585 + 142,895 144,480	738,259 − 589,767 147,492 + 589,767 738,259	685,978 − 498,369 187,609 + 498,369 685,978

Subtracting Numbers Written Horizontally

When the numbers you want to subtract are not in vertical columns, rewrite them with the larger number on top. Make sure that you line up the units under the units, the tens under the tens, and so on. Always line up the units column *first.*

EXAMPLE 7,522 − 971 =

REWRITE AS

$$\begin{array}{r} 7{,}522 \\ -\ 971 \\ \hline \end{array}$$

Subtract and check.

1. 7,522 − 971 = 9,330 − 827 = 5,942 − 307 =

2. 2,165 − 2,076 = 6,752 − 4,397 = 9,283 − 2,474 =

3. 3,275 − 2,486 = 5,680 − 4,597 = 7,223 − 6,445 =

4. 57,542 − 9,651 = 24,143 − 5,048 = 82,633 − 3,272 =

5. 21,456 − 3,569 = 26,116 − 5,248 = 53,224 − 4,097 =

6. 61,510 − 42,513 = 43,524 − 22,685 = 15,697 − 14,938 =

Subtracting from Zeros

You cannot borrow from zero. When the digit in the column you want to borrow from is zero, move to the next column to the left that does *not* contain a zero. Read through the steps of the next example carefully.

EXAMPLE 1

```
 3 1
 406
-159
```

STEP 1 You cannot subtract 9 from 6. You cannot borrow 1 from the zero in the tens column. Borrow 1 from the hundreds column and take it to the empty tens column. You now have 10 tens in the tens column. This leaves 3 hundreds.

```
 3 9
   1 1
 406
-159
```

STEP 2 Borrow 1 from the tens column and take it to the units column. You now have 10 + 6 = 16 in the units column. This leaves 9 tens.

```
 3 9
   1 1
 406
-159 } +
 247 }
 406 ✓
```

STEP 3 Subtract the units 16 − 9 = 7.
STEP 4 Subtract the tens 9 − 5 = 4.
STEP 5 Subtract the hundreds 3 − 1 = 2.
STEP 6 Check 159 + 247 = 406.

Subtract and check.

1.	502 − 263	308 − 149	604 − 326	305 − 278	207 − 199	401 − 285
2.	800 − 263	300 − 127	500 − 233	400 − 291	600 − 382	200 − 194
3.	3,004 − 973	5,006 − 425	2,007 − 347	4,003 − 621	9,002 − 972	6,001 − 430
4.	8,002 − 628	9,003 − 324	4,005 − 556	3,006 − 439	8,007 − 299	1,006 − 307
5.	4,003 − 2,564	6,002 − 3,574	3,008 − 2,569	9,004 − 2,916	5,005 − 4,936	8,003 − 1,227

EXAMPLE 2

$$\begin{array}{r} 6{,}000 \\ -\ 2{,}786 \\ \hline \end{array} \qquad \begin{array}{r} \overset{5}{\not{6}},\overset{1\,9}{\not{0}}\overset{1\,9}{\not{0}}\overset{1}{0} \\ -\ 2{,}786 \\ \hline 3{,}214 \\ \hline 6{,}000 \end{array}$$

EXAMPLE 3

$$\begin{array}{r} 60{,}704 \\ -\ 59{,}857 \\ \hline \end{array} \qquad \begin{array}{r} \overset{5}{\not{6}}\overset{1\,9}{\not{0}},\overset{1\,16}{\not{7}}\overset{1\,9}{\not{0}}\overset{1}{4} \\ -\ 59{,}857 \\ \hline 847 \\ \hline 60{,}704 \end{array}$$

Subtract and check.

6.	5,000 − 436	4,000 − 228	8,000 − 927	1,000 − 533	6,000 − 407	7,000 − 316
7.	9,000 − 2,544	7,000 − 1,633	2,000 − 1,429	4,000 − 2,338	3,000 − 1,426	5,000 − 3,258
8.	2,040 − 1,537	3,004 − 1,252	5,000 − 2,430	6,200 − 2,576	3,060 − 1,485	4,002 − 3,268
9.	20,000 − 18,954	40,000 − 25,364	30,000 − 11,456	50,000 − 23,812	70,000 − 42,307	
10.	40,007 − 23,564	80,006 − 51,288	70,003 − 29,436	10,004 − 9,543	50,002 − 23,875	

Rewrite the problems on this page with the larger number on top and the smaller number directly below it. Put units under units, tens under tens, and so on. In problems with money, put dollars under dollars and cents under cents.

1. 407 − 98 =	602 − 55 =	$9 − $0.27 =
2. 800 − 28 =	500 − 73 =	$3 − $0.44 =
3. 700 − 256 =	600 − 421 =	$4 − $1.09 =
4. 4,080 − 493 =	6,070 − 576 =	$20.50 − $2.88 =
5. 3,090 − 1,987 =	7,050 − 3,456 =	$60.20 − $20.71 =
6. 8,004 − 438 =	4,003 − 927 =	$90.02 − $6.05 =
7. 20,000 − 2,054 =	60,000 − 3,118 =	$400 − $64.17 =

Subtraction Shortcuts

When the subtrahend (the number you subtract from another number) ends in zero, you can subtract in your head. Any number minus zero is that number.

EXAMPLE 1 **46 − 30 =**

STEP 1 Do not rewrite the problem. You know that 6 − 0 = 6.

STEP 2 You also know that 40 − 30 = 10.

STEP 3 The answer to 46 − 30 = 16.

STEP 4 Check by adding 16 + 30 = 46.

Subtract each problem in your head.

1. 72 − 40 = 32 + 40 = 72 97 − 60 = 37 + 60 = 97 123 − 80 = 43 + 80 = 123

2. 38 − 10 = 28 + 10 = 38 41 − 30 = 11 + 30 = 41 156 − 90 = 66 + 90 = 156

3. 65 − 20 = 45 + 20 = 65 84 − 50 = 34 + 50 = 84 139 − 70 = 69 + 70 = 139

Sometimes you can change a subtraction problem to a similar problem that is easier to subtract. Look at the next example carefully.

EXAMPLE 2 **94 − 67 =**

STEP 1 Add 3 to both numbers to make the subtrahend end with zero.
94 − 67 =
+ 3 + 3

STEP 2 Subtract the new numbers. 97 − 70 = 27

STEP 3 Check by subtracting 67 from 94. 94 − 67 = 27

Rewrite each problem as a similar problem with a subtrahend that ends in zero. Remember to add to *both* numbers in the problem. Then subtract the new numbers.

4. 83 − 26 = 104 − 88 = 52 − 19 =

5. 92 − 45 = 71 − 35 = 114 − 79 =

6. 90 − 63 = 85 − 56 = 64 − 28 =

7. 96 − 38 = 85 − 17 = 131 − 47 =

Rounding and Estimating

Review the rules for rounding numbers on page 11 and the example on page 23. Use your skills in rounding numbers to estimate the answers to the problems on this page.

For problems 1 to 6, round each number to the nearest *hundred.* Then subtract the rounded numbers.

1. $42{,}356 - 9{,}485 \approx$

2. $38{,}182 - 8{,}491 \approx$

3. $87{,}421 - 9{,}530 \approx$

4. $78{,}253 - 9{,}682 \approx$

5. $34{,}818 - 5{,}847 \approx$

6. $92{,}309 - 8{,}958 \approx$

For problems 7 to 12, round each number to the nearest *thousand.* Then subtract the rounded numbers.

7. $22{,}483 - 19{,}584 \approx$

8. $62{,}537 - 56{,}849 \approx$

9. $55{,}130 - 43{,}273 \approx$

10. $25{,}614 - 16{,}726 \approx$

11. $678{,}054 - 97{,}347 \approx$

12. $327{,}583 - 58{,}946 \approx$

For problems 13 to 18, round each number to the nearest *ten thousand.* Then subtract the rounded numbers.

13. $208{,}625 - 120{,}432 \approx$

14. $524{,}088 - 418{,}950 \approx$

15. $809{,}307 - 326{,}175 \approx$

16. $473{,}162 - 213{,}076 \approx$

17. $398{,}416 - 106{,}432 \approx$

18. $189{,}268 - 98{,}416 \approx$

For problems 19 to 22, round each number to the nearest *hundred thousand.* Then subtract the rounded numbers.

19. $2{,}543{,}950 - 418{,}513 \approx$

20. $6{,}473{,}032 - 1{,}214{,}877 \approx$

21. $9{,}144{,}480 - 1{,}342{,}895 \approx$

22. $2{,}738{,}259 - 2{,}589{,}767 \approx$

Applying Your Subtraction Skills

The problems in this exercise require you to apply your subtraction skills to practical situations. In each problem, pay close attention to the language that tells you to subtract. Watch for words such as **difference, decrease, increase,** and **balance.** These words suggest subtraction. And watch for phrases such as **how much more, how much less, how much change,** and **how much is left.** These phrases also suggest subtraction.

Keep in mind that in subtraction problems you often have to *compare* two numbers.

After solving many of these problems, you will be asked to estimate your answers. Remember that estimation is a good way to check whether an answer is reasonable.

Solve and write the correct label, such as $ or miles, next to each answer.

1. In Phoenix, Arizona, the lowest temperature on record is 17°. The highest temperature on record is 122°. What is the difference between the highest and lowest temperatures?

2. To estimate the answer to the last problem, first round each temperature to the nearest *ten.* Then use the rounded numbers to find the difference.

3. Carlos paid $178.45 for uniforms for his new job at a car repair shop. How much change did he get from $200?

4. Estimate the amount of change Carlos received in the last problem by first rounding each amount to the nearest *ten* dollars.

5. In his baseball career, Pete Rose got a total of 4,256 hits. In his career, Ty Cobb got 4,191 hits. Rose got how many more hits than Cobb?

6. Estimate the difference between Pete Rose's total number of hits and Ty Cobb's total number of hits by first rounding each total to the nearest *ten.*

7. In Midvale, the average annual income for a family with two wage earners is $48,231. For families with only one wage earner, the average is $27,649. How much greater is the average income for a two-earner family than the average income for a one-earner family?

8. Estimate the answer to the last problem by first rounding each amount to the nearest *thousand* dollars.

9. The Mississippi River is 1,171 miles long. The Ohio River is 981 miles long. How much longer is the Mississippi River than the Ohio River?

10. Estimate the answer to the last problem by first rounding the length of each river to the nearest *ten* miles.

11. Mr. and Mrs. Nazir made a down payment of $20,175 on a house. The purchase price of the house was $134,500. The balance left after making the down payment is the amount of their mortgage. Find the amount of the mortgage.

12. To estimate the amount of the mortgage in the last problem, round the down payment and the purchase price to the nearest *thousand* dollars.

13. Suzanne is the chief rental agent for a new housing complex. Plans call for 1,284 apartments in the complex. By the time the buildings were completed, 497 of the apartments were already rented. How many apartments were still available?

14. Estimate the number of available apartments by first rounding the numbers to the nearest *hundred.*

15. Carla and Fred paid $86,990 for their house. Fifteen years later, a real estate agent told them that their house was worth $149,500. By how much did the price of the house increase in fifteen years?

16. Estimate the increase in the value of Carla and Fred's house by first rounding each price to the nearest *thousand* dollars.

17. Carmen works part-time as a dressmaker. Her goal this year is to make at least $20,000 at her part-time business. By the end of August, Carmen had earned $13,786. How much more does she have to make to reach her goal for the year?

18. Estimate the answer to the last problem by first rounding the amount Carmen has earned to the nearest *thousand* dollars.

The next four problems do not have the key words or phrases that suggest subtraction. These problems often ask you to compare amounts over a period of time. For example, you may be asked to compare costs as they change over time.

19. An inn in Pennsylvania first opened for business in 1782. The inn stayed in business over the years. For how many years had the inn been operating by the year 2010?

20. A mountain bike originally sold for $290. The bike is on sale for $198.80. How much can Chris save if he buys the bike on sale?

21. Sergio drives a truck four days a week. On Monday morning the odometer (mileage gauge) on his truck had a reading of 48,762. On Thursday night when he finished his work week, the odometer reading was 50,350. How many miles did Sergio drive his truck that week?

22. Anna retired from her job as a bookkeeper at the age of 78. When she retired, she realized that she had worked at the same company for 59 years. How old was Anna when she began working for the company?

23. For many years, experts believed that Mount Everest, the highest mountain in the world, was 29,028 feet high. Recently, scientists discovered that Mount Everest is in fact 29,035 feet high. The old height was wrong by how many feet?

24. The second highest mountain, K2, is 28,250 feet high. Using the most recent calculation for the height of Mount Everest, find how much higher Mount Everest is than K2.

25. According to the Census Bureau the population of the United States in 2010 was 310,233,000. Some experts predict that in 2030 the U.S. population will be 373,504,000. By how many people is the population expected to increase from 2010 to 2030?

26. Estimate the increase in the U.S. population from 2010 to 2030 by first rounding each number to the nearest *hundred thousand.*

27. Estimate the increase in the U.S. population from 2010 to 2030 by first rounding each number to the nearest *million.*

Multistep Problems

Most of the problems in this exercise require that you use both addition and subtraction. You have learned several key words and phrases that suggest either addition or subtraction. Read these problems carefully to be sure that you answer the question that is asked.

Solve and write the correct label, such as $ or cars, next to each answer.

1. Adnan teaches a math class on Tuesday and Thursday evenings. Usually there are 13 women in his class and 9 men. The classroom has 30 student seats. When all his students are in attendance, how many empty seats are there?

2. Mr. Dillon makes $28,970 a year working as a mechanic, and his wife makes $33,408 working as a kindergarten teacher. Their daughter Lisa is a college student who lives with her parents. Altogether the Dillons make $74,597 a year including Lisa's job at a fast-food restaurant. How much does Lisa make at her job?

Use the following information to answer questions 3 and 4.

The U.S. Congress is made up of the House of Representatives and the Senate. The following table shows the political party affiliation of the members of Congress in a recent year.

	Democrat	Republican	Other
Representatives	257	178	0
Senators	58	40	2

3. The total number of Democrats in Congress is how much greater than the total number of Republicans?

4. The total number of representatives is how much greater than the total number of senators?

5. Sharon has to book flights for people in her office who plan to attend a conference in Miami. Sharon went online to find round-trip airfare between New York and Miami. The best deal she found was an economy fare for $277 plus $21 in taxes and fees. The most expensive flight she found was a first-class fare for $953 plus $33 in taxes and fees. What is the difference between the most expensive fare and the least expensive fare including taxes and fees?

6. Joshua is a full-time college student. To pay his living expenses, Joshua works as a waiter on the weekends. Every month he hopes to earn $800 in tips and wages. The first weekend in May, Joshua made $168, the second weekend he made $233, and the third weekend he made $191. How much does Joshua have to make in tips and wages on the fourth weekend of May to reach his monthly goal?

7. Estimate the amount Joshua needs to make on the fourth weekend by rounding his earnings to the nearest *ten* dollars.

Use the following information to answer questions 8 and 9.

On a recent September 30, the three leading U.S. newspapers sold the following number of copies:

USA Today	2,293,310
The Wall Street Journal	2,011,999
The New York Times	1,000,665

8. Together, *The Wall Street Journal* and *The New York Times* sold how many more papers than *USA Today?*

9. Round each number to the nearest *thousand,* and estimate the combined circulation of the three leading U.S. newspapers.

10. The Rosa family takes home $3,475 every month. They try each month to live within the guidelines of a budget. For housing, they spend $1,040 each month. For transportation, they spend $560 each month. For food, they spend $695. How much do they have left over each month after paying for housing, transportation, and food?

Use the following information to answer questions 11 and 12.

This table shows the number of cars sold in the United States in a recent year. The number of domestic cars sold refers to cars made in the United States, Canada, and Mexico.

Domestic	4,535,098
Japan	1,141,768
Germany	506,736
Other countries	629,767

11. How many more domestic cars were sold than the combined number of cars produced in Japan, Germany, and other countries?

12. Estimate the total number of cars sold in the United States by first rounding the sales figures to the nearest *thousand.*

Subtraction Review

This review covers the material you have studied so far in this book. Read the signs (+ and −) carefully. When you finish, check your answers at the back of the book.

1. What is the value of the digit 8 in the number 1,839,567?

2. The population of Madison County is 249,538. What is the population rounded to the nearest *ten thousand?*

3. $\begin{array}{r} 873 \\ -\ 452 \\ \hline \end{array}$

4. $\begin{array}{r} 968 \\ -\ 202 \\ \hline \end{array}$

5. $\begin{array}{r} 637 \\ -\ \ 49 \\ \hline \end{array}$

6. $\begin{array}{r} 536 \\ -\ 287 \\ \hline \end{array}$

7. $\begin{array}{r} 4{,}850 \\ -\ 2{,}639 \\ \hline \end{array}$

8. $\begin{array}{r} 800 \\ -\ 492 \\ \hline \end{array}$

9. $\begin{array}{r} 7{,}002 \\ -\ 2{,}567 \\ \hline \end{array}$

10. $\begin{array}{r} 20{,}030 \\ -\ \ 4{,}038 \\ \hline \end{array}$

11. $\begin{array}{r} 300{,}000 \\ -\ 214{,}500 \\ \hline \end{array}$

12. 2,784 + 1,956 =

13. 61 + 86 + 49 + 52 =

14. 9,238 − 479 =

15. 406,000 − 209,513 =

16. 60,080 − 4,236 =

17. 70,005 − 38,028 =

18. Round each number to the nearest *thousand* and subtract.

348,214 − 198,247 ≈

19. Round each number to the nearest *ten thousand* and subtract.

894,328 − 59,069 ≈

Round each number to the nearest *hundred thousand* and subtract.

2,456,320 − 1,804,317 ≈

21. In a recent year 238,444 immigrants became permanent residents in California. In the same year 143,679 immigrants became permanent residents in New York. The number of new permanent residents in California was how much greater than the number in New York?

22. Ethel sold her car for $4,988. This was $1,437 less than she paid for the car when she bought it a year ago. How much did she pay for the car?

23. The Wongs bought new furniture for $630. If they made a down payment of $145.50, how much did they have left to pay?

Use the following information to answer questions 24 and 25.

24. In a recent year 476,469 of the most popular light truck were sold in the United States. In the same year 465,065 of the second-most popular light truck were sold. Round each number to the nearest *ten thousand* to estimate the combined sales of the most popular truck and the second-most popular.

25. Two years earlier, 744,996 of the most popular light truck were sold in the United States. Round the sales numbers to the nearest *thousand* to estimate the decrease in sales of the most popular light truck in two years.

SUBTRACTION REVIEW CHART

If you missed more than one problem on any group below, review the practice pages for those problems. Then redo the problems you got wrong before going on to the Multiplication Skills Inventory. If you had a passing score, redo any problem you missed and begin the Multiplication Skills Inventory on page 54.

Problem Numbers	Skill Area	Practice Pages
1	place value	6–10
2	rounding whole numbers	11–12
3, 4	subtraction facts	33–35
5, 6, 7	subtracting and borrowing	36–38
8, 9, 10, 11	subtracting from zeros	40–42
12, 13	adding horizontally	21
14, 15, 16, 17	subtracting horizontally	39, 42
18, 19, 20	rounding and estimating	44
21, 23, 25	applying subtraction	45–50
22, 24	applying addition	25–28

PLICATION

ication Skills Inventory

blems that you can. There is no time limit. Work carefully and check your answers, but do not use outside help. Correct answers are listed by page number at the back of the book.

1. $\begin{array}{r} 42 \\ \times\ 3 \\ \hline \end{array}$

2. $\begin{array}{r} 61 \\ \times\ 5 \\ \hline \end{array}$

3. $\begin{array}{r} 723 \\ \times\ 3 \\ \hline \end{array}$

4. $\begin{array}{r} 8{,}102 \\ \times\ 4 \\ \hline \end{array}$

5. $\begin{array}{r} 52 \\ \times\ 23 \\ \hline \end{array}$

6. $\begin{array}{r} 910 \\ \times\ 62 \\ \hline \end{array}$

7. $512 \times 34 =$

8. $6{,}122 \cdot 43 =$

9. $\begin{array}{r} 42 \\ \times\ 8 \\ \hline \end{array}$

10. $(236)(9) =$

11. $\begin{array}{r} 84 \\ \times\ 37 \\ \hline \end{array}$

12. $\begin{array}{r} 87 \\ \times\ 46 \\ \hline \end{array}$

13. $5{,}936(27) =$

14. $6{,}254 \cdot 83 =$

15. $(8{,}395)(516) =$

16. $92{,}047 \times 506 =$

17. $38{,}520 \cdot 780 =$

18. $47 \times 100 =$

19. $1{,}000(526) =$ **20.** $(4{,}936)(279) =$ **21.** $638 \times 5{,}076 =$

22. Anna runs 4 miles per day 364 days a year. Round 364 to the nearest ten and estimate the number of miles Anna runs in a year.

23. Fumio takes home $2,548 every month from his job installing kitchen cabinets. How much does he take home in one year?

24. A plane flies at an average speed of 413 miles per hour for 9 hours. How far does the plane fly in that time?

25. On a map of the United States, 1 inch represents 280 miles. How far apart are two cities that are 7 inches apart on the map?

MULTIPLICATION SKILLS INVENTORY CHART

If you missed more than one problem on any group below, work through the practice pages for that group. Then redo the problems you got wrong on the Multiplication Skills Inventory. If you had a passing score on all five groups of problems, redo any problem you missed and begin the Division Skills Inventory on page 82.

Problem Numbers	Skill Area	Practice Pages
1, 2, 3, 4	multiplication facts	56–60
5, 6, 7, 8	multiplying larger numbers	61–64
9, 10, 11, 12, 13, 14, 15, 16, 17, 18, 19, 20, 21	multiplying and carrying	65–69
22	rounding and estimating	70–72
23, 24, 25	applying multiplication	73–78

Basic Multiplication Facts

This page and the following two pages will help you learn the basic multiplication facts. The multiplication facts are basic building blocks for further study of mathematics. You must *memorize* these facts. Check your answers to this exercise. Practice the facts you got wrong. Then try these problems again until you can get all the answers quickly and accurately.

Parts of a Multiplication Problem

$$\begin{array}{r} 82 \\ \times \ 3 \\ \hline 246 \end{array}$$

82 ← multiplicand; × 3 ← multiplier — These numbers are also called **factors.**
246 ← product

1. $8 \times 4 =$	$1 \times 11 =$	$7 \times 8 =$	$3 \times 7 =$	$0 \times 8 =$
2. $5 \times 12 =$	$9 \times 2 =$	$7 \times 12 =$	$8 \times 2 =$	$2 \times 11 =$
3. $4 \times 5 =$	$9 \times 6 =$	$5 \times 6 =$	$6 \times 8 =$	$4 \times 11 =$
4. $4 \times 3 =$	$11 \times 12 =$	$6 \times 0 =$	$7 \times 7 =$	$1 \times 12 =$
5. $2 \times 12 =$	$8 \times 8 =$	$3 \times 4 =$	$3 \times 6 =$	$5 \times 8 =$
6. $9 \times 5 =$	$10 \times 12 =$	$11 \times 11 =$	$8 \times 6 =$	$6 \times 5 =$
7. $10 \times 11 =$	$9 \times 9 =$	$9 \times 4 =$	$6 \times 12 =$	$3 \times 11 =$
8. $6 \times 7 =$	$4 \times 4 =$	$4 \times 8 =$	$12 \times 11 =$	$3 \times 9 =$
9. $7 \times 9 =$	$7 \times 1 =$	$9 \times 11 =$	$8 \times 12 =$	$5 \times 11 =$
10. $7 \times 6 =$	$1 \times 5 =$	$4 \times 6 =$	$8 \times 3 =$	$3 \times 12 =$
11. $8 \times 9 =$	$5 \times 5 =$	$6 \times 6 =$	$4 \times 12 =$	$8 \times 11 =$
12. $8 \times 7 =$	$7 \times 4 =$	$9 \times 8 =$	$7 \times 11 =$	$10 \times 10 =$

13. $6 \times 7 =$	$9 \times 9 =$	$9 \times 5 =$	$10 \times 11 =$	$8 \times 8 =$
14. $11 \times 11 =$	$3 \times 6 =$	$1 \times 12 =$	$5 \times 8 =$	$9 \times 4 =$
15. $2 \times 12 =$	$5 \times 0 =$	$10 \times 12 =$	$7 \times 7 =$	$4 \times 3 =$
16. $6 \times 5 =$	$11 \times 12 =$	$5 \times 6 =$	$3 \times 4 =$	$6 \times 8 =$
17. $2 \times 11 =$	$4 \times 5 =$	$9 \times 2 =$	$4 \times 11 =$	$9 \times 6 =$
18. $7 \times 12 =$	$3 \times 7 =$	$8 \times 2 =$	$0 \times 3 =$	$8 \times 4 =$
19. $1 \times 11 =$	$5 \times 12 =$	$7 \times 8 =$	$9 \times 7 =$	$7 \times 3 =$
20. $6 \times 9 =$	$12 \times 12 =$	$6 \times 11 =$	$5 \times 9 =$	$9 \times 8 =$
21. $7 \times 5 =$	$9 \times 3 =$	$10 \times 10 =$	$9 \times 12 =$	$7 \times 11 =$
22. $8 \times 11 =$	$8 \times 7 =$	$5 \times 5 =$	$7 \times 4 =$	$6 \times 6 =$
23. $8 \times 3 =$	$4 \times 12 =$	$7 \times 6 =$	$8 \times 9 =$	$3 \times 12 =$
24. $1 \times 5 =$	$9 \times 11 =$	$4 \times 6 =$	$8 \times 12 =$	$3 \times 9 =$
25. $5 \times 11 =$	$7 \times 9 =$	$4 \times 4 =$	$6 \times 1 =$	$4 \times 8 =$
26. $6 \times 12 =$	$12 \times 11 =$	$3 \times 11 =$	$8 \times 6 =$	$9 \times 8 =$
27. $6 \times 9 =$	$5 \times 9 =$	$7 \times 5 =$	$9 \times 12 =$	$7 \times 3 =$
28. $9 \times 3 =$	$12 \times 12 =$	$9 \times 7 =$	$4 \times 7 =$	$6 \times 11 =$

29.	12 × 9	9 × 8	5 × 5	7 × 6	3 × 9	12 × 6	9 × 5
30.	12 × 1	4 × 3	11 × 2	3 × 7	11 × 6	10 × 10	12 × 4
31.	4 × 6	7 × 1	6 × 7	11 × 11	12 × 10	3 × 4	9 × 6
32.	12 × 5	7 × 3	11 × 7	6 × 6	1 × 5	7 × 9	11 × 3
33.	11 × 10	12 × 2	12 × 11	4 × 5	0 × 6	9 × 3	6 × 9
34.	11 × 8	8 × 3	11 × 9	4 × 4	8 × 6	8 × 8	9 × 0
35.	5 × 6	9 × 2	8 × 4	12 × 12	5 × 9	8 × 7	12 × 3
36.	12 × 8	9 × 9	3 × 6	7 × 7	6 × 8	12 × 7	11 × 1
37.	9 × 7	7 × 5	8 × 9	11 × 5	7 × 4	4 × 8	12 × 11
38.	9 × 4	5 × 8	6 × 5	11 × 4	8 × 2	7 × 8	9 × 12

The Multiplication Table

It is very important to know the multiplication table. If you do not know it, take the time to memorize it. The time you spend memorizing the table now will be saved later on because you will be able to do long multiplication and division problems quickly.

This table does not include 0 because there is only one thing to remember when you multiply by 0: any number multiplied by 0 is 0.

	1	2	3	4	5	6	7	8	9	10	11	12
1	1	2	3	4	5	6	7	8	9	10	11	12
2	2	4	6	8	10	12	14	16	18	20	22	24
3	3	6	9	12	15	18	21	24	27	30	33	36
4	4	8	12	16	20	24	28	32	36	40	44	48
5	5	10	15	20	25	30	35	40	45	50	55	60
6	6	12	18	24	30	36	42	48	54	60	66	72
7	7	14	21	28	35	42	49	56	63	70	77	84
8	8	16	24	32	40	48	56	64	72	80	88	96
9	9	18	27	36	45	54	63	72	81	90	99	108
10	10	20	30	40	50	60	70	80	90	100	110	120
11	11	22	33	44	55	66	77	88	99	110	121	132
12	12	24	36	48	60	72	84	96	108	120	132	144

Reading the Multiplication Table

Look at the row of numbers running across the top of the table, and then look at the column of numbers running along the left side. You can multiply any number in the top row by any number in the column on the left and find the answer inside the table. For example, to find out how much 8×7 is, locate the number 8 in the top row and the number 7 in the column on the left. Run your finger down the 8 column until you have reached the row marked 7. The answer is 56. You also could have found the answer by starting at the 8 on the left and moving your finger across the table until you reached the vertical column of figures marked 7 at the top. The multiplication table can be read both horizontally and vertically.

Notice also that the column of numbers under the 7 and the row of numbers next to the 7 increase by seven each time ($7 + 7 = 14$; $14 + 7 = 21$; $21 + 7 = 28$; and so on).

Multiplying by One-Digit Numbers

Multiply each digit in the top number by the bottom number. The answer should be written from right to left, starting with the product of the ones column, then the tens column, then the hundreds column, and so on.

EXAMPLE 1

$$\begin{array}{r} 243 \\ \times\ \ 2 \\ \hline 486 \end{array}$$

STEP 1 $2 \times 3 = 6$
STEP 2 $2 \times 4 = 8$
STEP 3 $2 \times 2 = 4$

Checking a multiplication problem is sometimes more difficult than actually working the problem. One method is to go over your steps to try to find any errors you might have made. Another more reliable method is to divide the answer you got by the bottom number in the problem. If you get the top number in the original problem as the answer to the division problem, your multiplication is correct. If you are not sure of your division at this point, don't worry about using this method of checking a multiplication problem. You will begin practicing division on page 84.

EXAMPLE 2

$$\begin{array}{r} 243 \\ \times\ \ 2 \\ \hline 486 \end{array}$$

CHECK

$$2\overline{)486} = 243$$

Multiply and check.

1. 71 × 9   81 × 7   62 × 4   43 × 3   91 × 8   90 × 3   20 × 8   70 × 7

2. 312 × 4   821 × 3   611 × 9   401 × 6   502 × 4   601 × 9   801 × 7

3. 610 × 7   420 × 3   510 × 8   110 × 9   700 × 5   900 × 7   200 × 6

4. 4,201 × 4 = 16,804   3,102 × 3 = 9,306   5,021 × 2 = 10,042   6,011 × 8 = 48,088   7,101 × 6 = 42,606   8,011 × 7 = 56,077

Multiplying by Larger Numbers

When you multiply by a 2-digit number, be sure to begin your answer (the first partial product) directly under the ones column of the numbers being multiplied. Begin the second partial product directly under the tens column of the numbers being multiplied and the first partial product. Always multiply from right to left.

EXAMPLE

$$\begin{array}{lr} & 51 \\ & \times\ 32 \\ \hline \text{partial} & 102 \\ \text{products} & 153 \\ \hline & 1{,}632 \end{array}$$

STEP 1 $2 \times 1 = 2$
STEP 2 $2 \times 5 = 10$
STEP 3 $3 \times 1 = 3$
STEP 4 $3 \times 5 = 15$
STEP 5 Add the partial products to get the final product.

Multiply and check.

1. 62×23 | 23×31 | 31×57 | 92×43 | 24×22 | 52×34 | 81×75

2. 42×44 | 71×89 | 51×63 | 43×21 | 72×41 | 91×66 | 61×58

3. 50×73 | 30×29 | 90×64 | 70×52 | 20×18 | 80×47 | 40×36

4. 901×56 | 510×68 | $5{,}111 \times 89$ | $9{,}121 \times 34$ | $7{,}122 \times 43$ | $91{,}001 \times 78$

When multiplying by zero, write the answer 0 directly under the 0 in the problem. Then multiply by the next digit in the bottom number and continue to the left. The first problem is done for you as an example.

5.

82	71	52	91	62	43	31
× 40	× 80	× 30	× 50	× 40	× 30	× 90
3,280						

6.

641	312	511	713	912	411
× 20	× 40	× 70	× 20	× 30	× 80

Notice where the product of the hundreds column begins in problems with a 3-digit multiplier. Remember each partial product starts under and one column to the left of the previous partial product.

EXAMPLE

hundreds column

```
   10,321
×     113
   30 963 }
  103 21  } partial
 1 032 1  } products
1,166,273
```

7.

31,021	80,011	60,112	10,220	71,011
× 213	× 497	× 314	× 123	× 856

8.

51,423	40,203	23,121	12,332	31,212
× 112	× 232	× 321	× 302	× 443

Multiplying Numbers Written Horizontally

To multiply numbers written horizontally, rewrite the problem vertically with the shorter number on the bottom. Whether you put the shorter number on the top or on the bottom, you will get the same answer. However, by putting the shorter number on the bottom, you will have fewer partial products to add.

EXAMPLE 312 × 24 =

REWRITE

```
   312            24
 ×  24         × 312
  1 248 }         48 }
  6 24  }        24  }  partial
  7,488         7 2  }  products
                7,488
```

Multiply and check.

1. 41 × 3 =	81 × 5 =	64 × 2 =	53 × 3 =
2. 741 × 2 =	323 × 3 =	611 × 5 =	421 × 4 =
3. 331 × 23 =	622 × 41 =	812 × 34 =	913 × 22 =
4. 703 × 33 =	97 × 801 =	601 × 38 =	76 × 610 =
5. 212 × 33 =	43 × 31 =	73 × 22 =	93 × 133 =

So far in this book you have seen the × sign to indicate multiplication. There are three other common ways to indicate multiplication. Each of the following means "five times six is equal to thirty."

EXAMPLE 1 $5 \times 6 = 30$

EXAMPLE 2 $5 \cdot 6 = 30$ The raised dot means to multiply.

EXAMPLE 3 $5(6) = 30$ A number next to another number in parentheses means to multiply.

EXAMPLE 4 $(5)(6) = 30$ Numbers in parentheses with no sign between them mean to multiply.

Notice that there is not a sign, such as + or −, in 5(6) or (5)(6). When you study algebra, you will learn that 5 + (6) means to add and that (5) + (6) also means to add.

6. $941 \cdot 20 =$ $713 \cdot 30 =$ $811 \cdot 80 =$ $912 \cdot 40 =$

7. $21(540) =$ $34(620) =$ $75(810) =$ $32(930) =$

8. $(33)(9{,}131) =$ $(3{,}421)(22) =$ $(21)(6{,}124) =$ $(7{,}123)(23) =$

9. $51{,}211 \cdot 32 =$ $24(70{,}212) =$ $(61{,}201)(31) =$

10. $40(2{,}021) =$ $31 \cdot 81{,}202 =$ $(51)(61{,}001) =$

11. $(60)(80{,}010) =$ $21(40{,}321) =$ $12 \cdot 43{,}223 =$

Multiplying and Carrying

Carrying in multiplication is very much like carrying in addition. Be sure to multiply *first* and *then* add the number being carried. Carrying is sometimes called **regrouping** or **renaming.**

EXAMPLE

$$\begin{array}{r} 49 \\ \times\ 63 \\ \hline 147 \\ 2\,94 \\ \hline 3{,}087 \end{array}$$

STEP 1	$3 \times 9 = 27$	Write the 7.	Carry the 2.
STEP 2	$3 \times 4 = 12$	$12 + 2 = 14$	Write the 14.
STEP 3	$6 \times 9 = 54$	Write the 4.	Carry the 5.
STEP 4	$6 \times 4 = 24$	$24 + 5 = 29$	Write the 29.
STEP 5	Add the partial products.		

Multiply and check.

1.	17 × 4	64 × 5	95 × 7	76 × 2	83 × 9	43 × 6	66 × 3	52 × 8
2.	34 × 5	85 × 2	43 × 8	18 × 7	27 × 3	93 × 4	54 × 6	48 × 9
3.	69 × 2	36 × 9	24 × 6	85 × 3	26 × 5	48 × 7	74 × 4	53 × 8
4.	53 × 5	22 × 6	84 × 3	66 × 2	75 × 9	93 × 7	86 × 4	47 × 8
5.	86 × 6	64 × 4	77 × 3	29 × 8	45 × 7	33 × 5	97 × 2	52 × 9
6.	69 × 6	26 × 8	84 × 3	55 × 5	73 × 7	46 × 9	74 × 4	18 × 2

7.	97 × 3	43 × 9	76 × 4	66 × 3	38 × 8	57 × 6	13 × 7	28 × 5
8.	37 × 5	54 × 8	46 × 3	73 × 9	93 × 4	62 × 7	27 × 2	85 × 6
9.	76 × 6	38 × 2	26 × 8	47 × 5	19 × 4	53 × 9	64 × 3	93 × 7
10.	42 × 7	66 × 5	58 × 9	73 × 8	85 × 3	14 × 4	26 × 6	39 × 2

Remember: Any number multiplied by 0 is 0, and you must always multiply before you carry. If you remember this, you won't be tricked by the next row of problems on this page.

EXAMPLE

$$\begin{array}{r} 403 \\ \times\ 9 \\ \hline 3{,}627 \end{array}$$

STEP 1	9 × 3 = 27	Write the 7.	Carry the 2.
STEP 2	9 × 0 = 0	0 + 2 = 2	Write the 2.
STEP 3	9 × 4 = 36		

11.	509 × 3	407 × 6	803 × 5	602 × 9	708 × 8	109 × 2	406 × 7
12.	237 × 5	198 × 4	724 × 7	193 × 6	825 × 8	773 × 4	294 × 7
13.	862 × 9	754 × 3	817 × 6	563 × 5	359 × 4	268 × 5	738 × 9

14. (36)(27) = (82)(58) = (65)(43) = (94)(29) =

15. 87 · 63 = 34 · 74 = 68 · 53 = 56 · 48 =

16. 26(59) = 55(62) = 86(37) = 39(42) =

17. 56 × 83 = 72 × 76 = 84 × 38 = 93 × 66 =

18. (27)(73) = (64)(29) = (75)(66) = (93)(88) =

Remember to take the shortcut. Write the number with fewer digits on the bottom.

19. 43(255) = 10,965 ✓ 56(367) = 20,552 ✓ 623(79) = 49,217 784(38) = 29,792 ✓

255 × 43 = 10,965 ✓ 367 × 56 = 20,552 ✓ 623 × 79 = 49,217 ✓ 784 × 38 = 29,792 ✓

20. 68 · 895 = 96 · 273 = 408 · 53 = 907 · 27 =

When you multiply dollars and cents, remember to put a decimal point in the answer to separate dollars from cents.

21. (46)($3.05) = ($6.04)(89) = (73)($8.05) = (24)($7.06) =

22. $91.73 × 63 = 27 × $42.18 = $75.16 × 54 = 73 × $43.29 =

23. 82($12.56) = 56($37.09) = $60.48(81) = $86.90(29) =

Multiplying by 10, 100, and 1,000

To multiply a number by 10, add a 0 to the right of the number.

EXAMPLE 1 $25 \times 10 = 250$ or $\begin{array}{r} 25 \\ \times\ 10 \\ \hline 250 \end{array}$

To multiply a number by 100, add two 0's to the right of the number.

EXAMPLE 2 $36 \times 100 = 3{,}600$ or $\begin{array}{r} 36 \\ \times\ 100 \\ \hline 3{,}600 \end{array}$

To multiply a number by 1,000, add three 0's to the right of the number.

EXAMPLE 3 $721 \times 1{,}000 = 721{,}000$ or $\begin{array}{r} 721 \\ \times\ 1{,}000 \\ \hline 721{,}000 \end{array}$

1. $\begin{array}{r} 47 \\ \times\ 10 \\ \hline \end{array}$ $\begin{array}{r} 53 \\ \times\ 10 \\ \hline \end{array}$ $\begin{array}{r} 92 \\ \times\ 10 \\ \hline \end{array}$ $\begin{array}{r} 36 \\ \times\ 10 \\ \hline \end{array}$ $\begin{array}{r} 71 \\ \times\ 10 \\ \hline \end{array}$ $\begin{array}{r} 28 \\ \times\ 10 \\ \hline \end{array}$ $\begin{array}{r} 65 \\ \times\ 10 \\ \hline \end{array}$

2. $\begin{array}{r} 368 \\ \times\ 10 \\ \hline \end{array}$ $\begin{array}{r} 866 \\ \times\ 10 \\ \hline \end{array}$ $\begin{array}{r} 761 \\ \times\ 10 \\ \hline \end{array}$ $\begin{array}{r} 946 \\ \times\ 10 \\ \hline \end{array}$ $\begin{array}{r} 479 \\ \times\ 10 \\ \hline \end{array}$ $\begin{array}{r} 261 \\ \times\ 10 \\ \hline \end{array}$

3. $\begin{array}{r} 9{,}483 \\ \times\ 100 \\ \hline \end{array}$ $\begin{array}{r} 2{,}356 \\ \times\ 100 \\ \hline \end{array}$ $\begin{array}{r} 3{,}079 \\ \times\ 100 \\ \hline \end{array}$ $\begin{array}{r} 4{,}308 \\ \times\ 100 \\ \hline \end{array}$ $\begin{array}{r} 5{,}570 \\ \times\ 100 \\ \hline \end{array}$ $\begin{array}{r} 6{,}090 \\ \times\ 100 \\ \hline \end{array}$

4. $\begin{array}{r} 100 \\ \times\ 74 \\ \hline \end{array}$ $\begin{array}{r} 100 \\ \times\ 8 \\ \hline \end{array}$ $\begin{array}{r} 100 \\ \times\ 256 \\ \hline \end{array}$ $\begin{array}{r} 100 \\ \times\ 31 \\ \hline \end{array}$ $\begin{array}{r} 100 \\ \times\ 209 \\ \hline \end{array}$ $\begin{array}{r} 100 \\ \times\ 68 \\ \hline \end{array}$

5. $\begin{array}{r} 1{,}000 \\ \times\ 73 \\ \hline \end{array}$ $\begin{array}{r} 1{,}000 \\ \times\ 421 \\ \hline \end{array}$ $\begin{array}{r} 1{,}000 \\ \times\ 16 \\ \hline \end{array}$ $\begin{array}{r} 1{,}000 \\ \times\ 208 \\ \hline \end{array}$ $\begin{array}{r} 1{,}000 \\ \times\ 450 \\ \hline \end{array}$ $\begin{array}{r} 1{,}000 \\ \times\ 623 \\ \hline \end{array}$

6. 74 × 10 = | 31 × 10 = | 25 × 10 = | 56 × 10 =

7. 10 × 98 = | 10 × 46 = | 10 × 50 = | 10 × 37 =

8. 662 × 10 = | 296 × 10 = | 802 × 10 = | 429 × 10 =

9. 506 × 10 = | 380 × 10 = | 409 × 10 = | 110 × 10 =

10. 10 × 128 = | 10 × 839 = | 10 × 756 = | 10 × 217 =

11. 6,578 × 10 = | 7,308 × 10 = | 8,815 × 10 = | 7,049 × 10 =

12. 26 × 100 = | 43 × 100 = | 94 × 100 = | 37 × 100 =

13. 100 × 95 = | 100 × 30 = | 100 × 63 = | 100 × 74 =

14. 100 × 81 = | 100 × 57 = | 100 × 68 = | 100 × 25 =

15. 957 × 100 = | 214 × 100 = | 693 × 100 = | 898 × 100 =

16. 581 × 100 = | 142 × 100 = | 813 × 100 = | 420 × 100 =

17. 100 × 225 = | 100 × 730 = | 100 × 207 = | 100 × 396 =

18. 1,000 × 65 = | 1,000 × 88 = | 1,000 × 62 = | 1,000 × 91 =

19. 32 × 1,000 = | 603 × 1,000 = | 540 × 1,000 = | 931 × 1,000 =

20. 1,000 × 46 = | 1,000 × 261 = | 1,000 × 380 = | 1,000 × 715 =

21. 425 × 1,000 = 425,000 ✓ | 1,000 × 689 = 689,000 ✓ | 237 × 1,000 = 237,000 ✓ | 1,000 × 499 = 499,000 ✓

Rounding and Estimating

In the last exercise, you learned to multiply numbers by 10, 100, and 1,000 in your head. Zeros make many multiplication problems easier.

EXAMPLE 1 $6 \times 40 =$

STEP 1 Multiply $6 \times 4 = 24$.

$6 \times 40 = 240$

STEP 2 Bring along the 0 from 40.

EXAMPLE 2 $80(700) =$

STEP 1 Multiply $8 \times 7 = 56$.

$80(700) = 56{,}000$

STEP 2 Bring along the zeros from 80 and 700.

Multiply each problem in your head.

1. $40 \times 8 =$ $900 \times 60 =$ $(1{,}200)(30) =$

2. $9(70) =$ $5(700) =$ $9 \cdot 6{,}000 =$

3. $(60)(5) =$ $(20)(800) =$ $(2)(13{,}000) =$

4. $70 \cdot 20 =$ $5 \times 400 =$ $7{,}000 \times 80 =$

To estimate an answer to a multiplication problem, try rounding the larger number to the *left-most* place. Sometimes this is called **front-end rounding.**

For example, if the larger number in a problem is 427, round 427 to the nearest hundred. If the larger number is 2,846, round 2,846 to the nearest thousand.

EXAMPLE 3 Estimate the answer to 9×427.

$9 \times 427 \approx 9 \times 400$

STEP 1 Round 427 to the nearest hundred. Remember the symbol $\approx$ means "is approximately equal to."

$9 \times 400 = 3{,}600$

STEP 2 Multiply 9×400 in your head.

EXAMPLE 4 Estimate the answer to $5 \times 2{,}846$.

$5 \times 2{,}846 \approx 5 \times 3{,}000$

STEP 1 Round 2,846 to the nearest thousand.

$5 \times 3{,}000 = 15{,}000$

STEP 2 Multiply $5 \times 3{,}000$ in your head.

Round the larger number in each problem to the nearest *hundred* and multiply.

5. $4 \times 782 \approx$ $7(284) \approx$ $3(447) \approx$

6. $(912)(3) \approx$ $609 \times 2 \approx$ $8 \cdot 931 \approx$

7. $6 \cdot 472 \approx$ $(5)(872) \approx$ $189 \times 4 \approx$

Round the larger number in each problem to the nearest *thousand* and multiply.

8. $2(4{,}281) \approx$ $(7{,}516)(4) \approx$ $4(5{,}693) \approx$

9. $7 \times 2{,}963 \approx$ $5 \cdot 3{,}772 \approx$ $8 \cdot 1{,}204 \approx$

10. $3 \cdot 6{,}059 \approx$ $9{,}461 \times 6 \approx$ $(2)(18{,}366) \approx$

To get a quick estimate for an answer, you can use front-end rounding for both numbers.

EXAMPLE 5 Use front-end rounding to estimate the answer to 62×891.

$62 \times 891 \approx 60 \times 900$ — **STEP 1** Round 62 to the nearest ten and 891 to the nearest hundred.

$60 \times 900 = 54{,}000$ — **STEP 2** Multiply 60×900 in your head.

In the next problems, round the left-most digit in *both* numbers and multiply the rounded numbers in your head. The first problem is started for you.

11. $28 \cdot 73 \approx 30 \cdot 70 =$ $57(243) \approx$ $(79)(4{,}123) \approx$

12. $12 \times 294 \approx$ $88 \times 32 \approx$ $231 \times 659 \approx$

13. $4{,}809 \cdot 71 \approx$ $726 \cdot 16 \approx$ $428 \cdot 973 \approx$

In the following problems, use rounding to select the correct answer from the choices.

14. $94 \times 81 =$

a. 724
b. 7,614
c. 14,084
d. 17,234

15. $4{,}156 \times 18 =$

a. 41,208
b. 56,608
c. 74,808
d. 748,080

16. $56(73) =$

a. 4,088
b. 8,208
c. 14,084
d. 14,518

17. $(39)(6{,}234) =$

a. 18,646
b. 24,316
c. 186,246
d. 243,126

18. $36 \cdot 425 =$

a. 8,700
b. 9,600
c. 15,300
d. 21,900

19. $26 \times 5{,}085 =$

a. 132,210
b. 112,910
c. 90,670
d. 88,560

20. $(78)(112) =$

a. 3,226
b. 4,216
c. 6,106
d. 8,736

21. $5{,}736 \cdot 63 =$

a. 442,828
b. 361,368
c. 243,118
d. 186,208

22. $963 \times 48 =$

a. 28,804
b. 37,514
c. 39,084
d. 46,224

23. $(472)(681) =$

a. 152,642
b. 181,822
c. 243,512
d. 321,432

24. $21(734) =$

a. 9,364
b. 15,414
c. 18,704
d. 20,064

25. $798 \cdot 657 =$

a. 642,826
b. 524,286
c. 414,346
d. 382,156

Applying Your Multiplication Skills

The problems in this exercise require that you apply your multiplication skills to practical situations. In each problem, pay close attention to the language that tells you to multiply. In most cases, you will be *given information about one thing,* and you will be asked to *apply the information to several things.*

For example, you may be told how much someone makes in one hour and asked to find how much she makes in 40 hours. Or you may be told how many miles a driver can travel on one gallon of gasoline and asked how far the driver can travel on 12 gallons of gasoline.

After several of the problems in this exercise, you will be asked to estimate answers to check whether your answers are reasonable.

Solve and write the correct label, such as miles or gallons, next to each answer.

1. On a trip to visit her family, Maria drove her car at an average speed of 53 miles per hour. How far did she drive in four hours?

2. Estimate the distance Maria drove in the last problem by first rounding her average speed to the nearest *ten.*

3. On a regional map, one inch represents 68 miles. How many miles apart are two cities that are three inches apart on the map?

4. Estimate the distance between the two cities in the last problem by rounding the distance that one inch represents to the nearest *ten* miles.

5. The Uptown Theater has 43 rows of seats. Each row contains 58 seats. How many seats are there in the theater?

6. Round the number of rows and the number of seats in the last problem to the nearest *ten.* Then estimate the number of seats in the theater.

7. Steve had to buy four new tires for his truck. The tires each cost $139.99. How much did he pay for the new tires?

8. Estimate the price of Steve's tires by first rounding the cost of each tire to the nearest *ten* dollars.

9. Vera works in the data processing department of a large law office. She can type an average of 62 words per minute. How many words can she type in 25 minutes?

10. Use front-end rounding to estimate the number of words Vera can type in 25 minutes based on the last problem.

11. At Saturday night's performance at the local community college, 417 people each paid $18 to watch a talent show. What was the total value of the tickets that were sold?

12. Use front-end rounding to estimate the ticket sales in the last problem.

13. Joanne's new car gets about 27 miles per gallon when she drives on the highway. If the tank of her car holds 12 gallons, how far can she drive on a highway with a full tank of gasoline?

14. Use front-end rounding to estimate the distance Joanne can drive.

15. It costs $5,290 a year to educate a student at Oakdale School. What is the total yearly cost of educating 680 students at the school?

16. Use front-end rounding to estimate the cost of educating the students at Oakdale School for a year.

17. Helena owns a small convenience store. She makes a profit of $0.07 on every 16-ounce can of juice that she sells. If she sold 583 cans of juice last week, how much profit did she make on those sales?

18. Estimate Helena's profit by first rounding the number of cans to the nearest *hundred.*

19. Rashid bought a new computer. He has to pay $98.90 a month for 12 months. Round his monthly payment to the nearest *ten* dollars to estimate the total amount he will pay for the computer.

20. Guillermo lost his job in construction. He lives in New Mexico where he can get a weekly unemployment check for $278. How much will he get from the state if he is out of work for 13 weeks?

21. Use front-end rounding to estimate the total amount Guillermo will get from the state.

22. Each classroom at Midvale Central School holds 33 students. There are 19 classrooms in the school. How many students can Midvale Central enroll?

23. Use front-end rounding to estimate the total number of students the school can enroll.

24. A train travels at an average speed of 47 miles per hour. How far can the train travel in 12 hours?

25. Use front-end rounding to estimate the total distance the train can travel.

26. The sound of a lightning bolt striking a tree took five seconds to reach a listener. Sound travels at a speed of 1,129 feet per second. Round the speed of sound to the nearest *hundred* to estimate the distance between the tree that was struck and the listener.

Multistep Problems

Most of the problems in this exercise require that you use addition or subtraction as well as multiplication. Read these problems carefully to be sure that you answer the question that is asked.

Solve and write the correct label, such as $ or miles, next to each answer.

1. Alvaro makes $14.60 an hour at a factory that produces small electric motors. His normal workweek is 40 hours. For overtime work, Alvaro makes $21.90 an hour. One week when Alvaro's company had a large order to complete, he worked a total of 47 hours. How much did he earn that week?

2. Sam works in the storeroom of a grocery store. On Monday morning, he received 10 cases of cola, 8 cases of orange drink, and 6 cases of root beer from one supplier. Each case contains 24 cans. How many cans were in the entire delivery?

3. To pay off their mortgage, Christine and Jaime agree to pay $540 a month every month for 20 years. What total amount will they pay over 20 years?

4. Doreen wants to buy new living room furniture that costs $2,059. A salesperson at the store told Doreen that she could pay $129 a month for 24 months. How much more than the asking price will Doreen pay if she makes monthly payments?

5. Each building in Roosevelt Complex has 16 floors of apartments, and there are 12 apartments on each floor. Altogether there are seven buildings in the Roosevelt Complex. How many apartments are there in the complex?

6. The typical family living in Roosevelt Complex has four people. Which of the following is closest to the total number of residents in the complex?

   a. 520
   b. 1,040
   c. 5,200
   d. 10,400

7. To build an addition to a community center, six carpenters each worked 160 hours. The average pay for each carpenter was $24 an hour. Altogether how much did the community have to pay for the carpenters' wages?

8. To pay for a new side-by-side refrigerator/freezer, José and Linda agreed to make a down payment of $200 and then make 18 monthly payments of $129. What total price will they pay for the refrigerator/freezer?

9. The Meyer family left early Saturday morning to get to their cottage in the mountains. They drove for an hour in the city at an average speed of 12 mph. Then they drove three hours on a highway at an average speed of 65 mph. Finally they drove one more hour on country roads at an average speed of 40 mph. What total distance did they drive?

10. Martin owns a clothing store. He has to pay a supplier $54.60 for a popular style of jacket. He charges his customers $69.95 for the jacket. How much profit will Martin make if he sells 20 jackets?

11. Sally was in charge of ticket sales for a band concert. For a Friday night performance, she sold 328 tickets at $25 each, 119 senior discount tickets at $20 each, and 256 student discount tickets at $18 each. What was the total amount of the ticket sales for that performance?

12. A volunteer committee is trying to raise $1,500 to improve the playground in the neighborhood park. So far 38 families have given an average of $30 each. How much more money does the committee need to raise?

13. Roberto owns a grocery store. He pays his supplier $0.59 for an 11-ounce can of corn. He charges his customers $0.68 for a can of corn. The difference between the price Roberto charges his customers and the price he pays his supplier is called the markup. What is the total markup on 1,200 cans of corn?

14. Robin runs two machines at Precision Plastics. One machine can make 80 containers in an hour. The other machine can make 115 smaller containers in an hour. Altogether how many containers can Robin's machines make during her 8-hour shift?

Multiplication Review

This review covers the material you have studied so far in this book. Read the signs carefully to decide whether to add, subtract, or multiply. When you finish, check your answers at the back of the book.

1. Jake made $9,482 at his part-time job last year. What was his income to the nearest *hundred* dollars? $9,500 ✓

2. What is the value of the 9 in $9,482? $9,000

3. 14 + 37 + 6 = 57 ✓

4. 98,208 + 7,065 = 105,273 ✓

5. 23,584 − 12,270 = 11,314 ✓

pay attention to the substract!

6. 340,600 − 91,837 = 248,763 ✓

7. 62 × 4 = 248 ✓

8. 421 × 3 = 1,263 ✓

9. 610 × 58 = 35,380 ✓

10. 411 × 36 = 14,796 ✓

11. (73)(6) = 438 ✓

12. 8(457) = 3,656 ✓

13. 85 · 49 = 4,165 ✓

14. 736 · 53 = 39,008 ✓

15. 83,056 × 47 = HELP!!

16. 53 × 1,000 = 53,000 ✓

17. 100(423) = 42,300 ✓

18. 7,240 · 10 = 72,400 ✓

19. Round both numbers to the nearest *hundred* and multiply.

$(428)(782) \approx$

20. Round the smaller number to the nearest *hundred* and the larger number to the nearest *thousand.* Then multiply.

$469 \times 6{,}738 \approx$

21. Round the smaller number to the nearest *hundred* and the larger number to the nearest *ten thousand.* Then multiply.

$(860)(49{,}630) \approx$

22. Mr. and Mrs. Burns bought a used car. They agreed to pay $225 every month for 28 months. Which of the following is closest to the total amount of their payments?

a. $2,700
b. $3,600
c. $4,400
d. $6,000

23. There are 1,000 meters in one kilometer. How many meters are there in 23 kilometers?

24. If the average adult sleeps 7 hours a night, how many hours of sleep does he or she get in one year? (1 year = 365 days)

25. Before he took a speed reading course, Joaquin read 250 words per minute. When he finished the course, he was able to read 375 words per minute. By how many words per minute did his reading speed increase?

MULTIPLICATION REVIEW CHART

If you missed more than one problem on any group below, review the practice pages for those problems. Then redo the problems you got wrong before going on to the Division Skills Inventory. If you had a passing score, redo any problem you missed and begin the Division Skills Inventory on page 82.

Problem Numbers	Skill Area	Practice Pages
1, 2	place value	6–12
3, 4	addition	15–28
5, 6, 25	subtraction	33–50
7, 8, 9, 10	simple multiplication	56–64
11, 12, 13, 14, 15	multiplying and carrying	65–67
16, 17, 18	multiplying by 10, 100, 1,000	68–69
19, 20, 21	rounding and estimating	70–72
22, 23, 24	applying multiplication	73–78

N

Skills Inventory

you can. There is no time limit. Work carefully and check ... use outside help. Correct answers are listed by page number at the back of the book.

1. $8\overline{)184}$ **2.** $4\overline{)268}$ **3.** $9\overline{)2{,}142}$

4. $7\overline{)2{,}931}$ **5.** $6\overline{)4{,}225}$ **6.** $5\overline{)24{,}014}$

7. $371 \div 53 =$ **8.** $331 \div 62 =$ **9.** $\frac{432}{48} =$

10. $\frac{2{,}656}{32} =$ **11.** $\frac{6{,}663}{73} =$ **12.** $51{,}801 \div 83 =$

13. $24{,}882 / 429 =$ **14.** $40{,}837 \div 583 =$ **15.** $\frac{5{,}600}{80} =$

16. Round 2,763 to the nearest *hundred.* Then divide by 7.

$2{,}763 \div 7 \approx$

17. Round 18,274 to the nearest *thousand.* Then divide by 6.

$18{,}274 \div 6 \approx$

18. Which of the following is *closest* to the correct answer for 22,698 ÷ 39?

a. 390
b. 450
c. 580
d. 630

19. If a train travels at an average speed of 37 miles per hour, how much time will it take to go 814 miles?

20. A large carton weighs 2,608 ounces. There are 16 ounces in one pound. Find the weight of the carton in pounds.

21. The Martinez family pays $7,872 in rent in one year. How much rent does the family pay each month?

22. Petra bought four pairs of boots for her children. She paid $159.20 for all the boots. Find the average price for each pair of boots.

DIVISION SKILLS INVENTORY CHART

If you missed more than one problem on any group below, work through the practice pages for that group. Then redo the problems you got wrong on the Division Skills Inventory. If you had a passing score on all seven groups of problems, redo any problem you missed and go to Posttest A on page 114.

Problem Numbers	Skill Area	Practice Pages
1, 2, 3	dividing by 1 digit	87–89
4, 5, 6	dividing with remainders	90–92
7, 8, 9, 10, 11, 12	dividing by 2 digits	94–97
13, 14	dividing by 3 digits	98–99
15, 16, 17	rounding and estimating	100–101
18	2-digit accuracy	102–103
19, 20, 21, 22	applying division	104–110

Basic Division Facts

This page and the following page will help you learn the basic division facts. The division facts are basic building blocks for further study of mathematics. You must *memorize* these facts. Check your answers to this exercise. Practice the facts you got wrong. Then try these problems again until you can get all the answers quickly and accurately. If you have trouble, go back to the multiplication table on page 59 for review.

Parts of a Division Problem

There are four common ways to write division. The number being divided is called the **dividend.** The number that divides into the dividend is the **divisor.** The answer is the **quotient.** In each example below, the dividend is 20, the divisor is 4, and the quotient is 5.

$4\overline{)20}$ with quotient 5 $20 \div 4 = 5$ $\frac{20}{4} = 5$ $20 / 4 = 5$

1. $27 \div 9 =$ 3 $54 \div 6 =$ 9 $25 \div 5 =$ 5

2. $42 \div 6 =$ 7 $24 \div 8 =$ 3 $48 \div 8 =$ 6

3. $16 \div 4 =$ 4 $45 \div 5 =$ 9 $12 \div 4 =$ 3

4. $\frac{48}{6} =$ 8 $\frac{64}{8} =$ 8 $\frac{30}{5} =$ 6

5. $\frac{18}{2} =$ 9 $\frac{108}{12} =$ 9 $\frac{32}{4} =$ 8

6. $\frac{20}{5} =$ 4 $\frac{49}{7} =$ 7 $\frac{18}{6} =$ 3

7. $81 / 9 =$ 9 $63 / 9 =$ 7 $56 / 8 =$ 7

8. $36 / 6 =$ 6 $72 / 12 =$ 6 $21 / 7 =$ 3

9. $56 / 7 =$ 8 $88 / 11 =$ 8 $54 / 9 =$ 6

10. $15 \div 3 = 5$ | $80 \div 10 = 8$ | $40 \div 5 = 8$

11. $36 \div 4 = 9$ | $42 \div 7 = 6$ | $24 \div 6 = 7$

12. $72 \div 8 = 9$ | $28 \div 7 = 4$ | $35 \div 5 = 7$

13. $\frac{63}{7} = 9$ | $\frac{50}{5} = 10$ | $\frac{24}{3} = 8$

14. $\frac{16}{8} = 2$ | $\frac{22}{2} = 11$ | $\frac{32}{8} = 4$

15. $\frac{55}{11} = 5$ | $\frac{40}{8} = 5$ | $\frac{18}{3} = 6$

16. $60 / 12 = 5$ | $0 / 2 = 0$ | $45 / 9 = 5$

17. $4 / 1 = 4$ | $21 / 3 = 7$ | $60 / 5 = 12$

18. $30 / 6 = 5$ | $90 / 9 = 10$ | $24 / 12 = 2$

19. $84 \div 7 = 12$ | $33 \div 11 = 3$ | $14 \div 2 = 7$

20. $6 \div 3 = 2$ | $24 \div 4 = 6$ | $35 \div 7 =$

21. $72 \div 6 =$ | $8 \div 1 =$ | $9 \div 3 =$

22. $\frac{16}{2} =$ | $\frac{96}{8} =$ | $\frac{10}{5} =$

23. $\frac{48}{12} =$ | $\frac{14}{7} =$ | $\frac{12}{2} =$

24. $\frac{18}{9} =$ | $\frac{12}{3} =$ | $\frac{84}{12} =$

20 Questions
x 5 points
100 total Test

25. $8 / 4 =$ $\quad$ $28 / 4 =$ $\quad$ $9 / 1 =$

26. $27 / 3 =$ $\quad$ $0 / 6 =$ $\quad$ $88 / 8 =$

27. $36 / 9 =$ $\quad$ $30 / 10 =$ $\quad$ $8 / 2 =$

28. $15 \div 5 =$ $\quad$ $121 \div 11 =$ $\quad$ $60 \div 6 =$

29. $50 \div 10 =$ $\quad$ $40 \div 4 =$ $\quad$ $96 \div 12 =$

30. $7 \div 1 =$ $\quad$ $72 \div 9 =$ $\quad$ $0 \div 8 =$

31. $\frac{20}{5} =$ $\quad$ $\frac{66}{6} =$ $\quad$ $\frac{90}{10} =$

32. $\frac{22}{11} =$ $\quad$ $\frac{6}{1} =$ $\quad$ $\frac{8}{2} =$

33. $\frac{36}{3} =$ $\quad$ $\frac{108}{9} =$ $\quad$ $\frac{100}{10} =$

34. $40 / 10 =$ $\quad$ $36 / 12 =$ $\quad$ $33 / 3 =$

35. $20 / 10 =$ $\quad$ $132 / 12 =$ $\quad$ $44 / 4 =$

36. $6 / 2 =$ $\quad$ $12 / 6 =$ $\quad$ $66 / 11 =$

37. $99 \div 11 =$ $\quad$ $20 \div 2 =$ $\quad$ $60 \div 10 =$

38. $110 \div 11 =$ $\quad$ $70 \div 10 =$ $\quad$ $24 \div 2 =$

39. $144 \div 12 =$ $\quad$ $77 \div 11 =$ $\quad$ $10 \div 2 =$

Dividing by One-Digit Numbers

Look carefully at the example below to see how to do a division problem.

EXAMPLE

$$4\overline{)156}\ \text{quotient } 3$$

STEP 1 Divide 15 ÷ 4 = 3. Put the 3 over the 5 in the dividend.

$$\begin{array}{r} 3 \\ 4\overline{)156} \\ 12 \end{array}$$

STEP 2 Multiply 3 × 4 = 12. Put the 12 directly under the 15.

$$\begin{array}{r} 3 \\ 4\overline{)156} \\ \underline{12} \\ 3 \end{array}$$

STEP 3 Subtract 15 − 12 = 3.

$$\begin{array}{r} 3 \\ 4\overline{)156} \\ \underline{12} \\ 36 \end{array}$$

STEP 4 Bring down the next number from the dividend.

$$\begin{array}{r} 39 \\ 4\overline{)156} \\ \underline{12} \\ 36 \end{array}$$

STEP 5 Divide 36 ÷ 4 = 9. Put the 9 directly over the 6 in the dividend.

$$\begin{array}{r} 39 \\ 4\overline{)156} \\ \underline{12} \\ 36 \\ 36 \end{array}$$

STEP 6 Multiply 9 × 4 = 36. Put the 36 directly under the 36 in the problem.

$$\begin{array}{r} 39 \\ 4\overline{)156} \\ \underline{12} \\ 36 \\ \underline{36} \\ 0 \end{array}$$

STEP 7 Subtract 36 − 36 = 0.

When dividing by a 1-digit number, most of the work can be done by multiplying and subtracting mentally. Write each digit that you carry in the dividend. This procedure is called **short division.**

EXAMPLES

$$\begin{array}{r} 39 \\ 4\overline{)15^{3}6} \end{array} \qquad \begin{array}{r} 274 \\ 5\overline{)1{,}3^{3}7^{2}0} \end{array} \qquad \begin{array}{r} 4{,}867 \\ 2\overline{)9{,}^{1}7^{1}3^{1}4} \end{array}$$

To check a division problem, multiply the quotient by the divisor. The product should be the dividend. For the examples above:

$$\begin{array}{r} 39 \\ \times\ 4 \\ \hline 156 \end{array} \qquad \begin{array}{r} 274 \\ \times\ 5 \\ \hline 1{,}370 \end{array} \qquad \begin{array}{r} 4{,}867 \\ \times\ 2 \\ \hline 9{,}734 \end{array}$$

Divide and check.

1. 6)252   4)356   8)424   7)322   3)237

2. 3)162   9)801   2)126   6)504   8)688

3. 5)215   6)408   3)195   9)603   2)194

4. 8)272   5)360   4)264   7)336   6)174

5. 5)340   4)192   7)413   3)261   9)540

Rewrite each problem and divide. The first problem is started for you.

6. $360 \div 4 =$	$480 \div 8 =$	$240 \div 6 =$	$210 \div 3 =$	$450 \div 5 =$

$4\overline{)360}$

7. $2{,}768 \div 8 =$	$2{,}051 \div 7 =$	$1{,}940 \div 5 =$	$4{,}068 \div 9 =$
8. $2{,}325 \div 3 =$	$4{,}571 \div 7 =$	$2{,}065 \div 5 =$	$5{,}728 \div 8 =$
9. $\frac{3{,}624}{6} =$	$\frac{1{,}228}{4} =$	$\frac{7{,}232}{8} =$	$\frac{3{,}040}{5} =$
10. $1{,}418 / 2 =$	$3{,}448 / 8 =$	$2{,}468 / 4 =$	$3{,}432 / 6 =$

When you divide dollars and cents, remember to put a decimal point in the answer to separate dollars from cents.

11. $\frac{\$34.20}{9} =$ \$3.80	$\frac{\$33.05}{5} =$ \$33.05	$\frac{\$21.84}{7} =$ \$21.84	$\frac{\$20.32}{4} =$ \$5.80
12. $\$26.24 / 4 =$ \$6.56	$\$17.46 / 6 =$ \$2.91	X $\$18.27 / 9 =$ \$2.03	$\$37.52 / 7 =$ \$37.52
13. $\frac{\$22.20}{3} =$	$\frac{\$17.68}{8} =$	$\frac{\$23.70}{2} =$	$\frac{\$88.74}{9} =$

Dividing with Remainders

Division problems do not always come out evenly. The amount left over is called the **remainder.**

To check a division problem with a remainder, multiply the quotient by the divisor. Then add the remainder. The result should equal the dividend.

EXAMPLE

$$\begin{array}{r} 487 \text{ r } 4 \\ 5\overline{)2{,}439} \\ \underline{20} \\ 43 \\ \underline{40} \\ 39 \\ \underline{35} \\ 4 \end{array}$$

CHECK

$$\begin{array}{r} 487 \\ \underline{\times \quad 5} \\ 2{,}435 \\ \underline{+ \quad\quad 4} \\ 2{,}439 \end{array}$$

Divide and check.

1. $6\overline{)1{,}449}$ $8\overline{)5{,}629}$ $5\overline{)2{,}352}$ $7\overline{)2{,}059}$

2. $4\overline{)2{,}431}$ $6\overline{)2{,}495}$ $3\overline{)1{,}537}$ $9\overline{)8{,}197}$

3. $2\overline{)1{,}619}$ $5\overline{)1{,}842}$ $7\overline{)2{,}946}$ $4\overline{)2{,}867}$

Rewrite each problem. Then divide and check.

4. 4,295 ÷ 8 = 2,835 ÷ 4 = 5,954 ÷ 9 = 1,259 ÷ 3 =

5. 1,321 ÷ 7 = 1,618 ÷ 6 = 2,163 ÷ 5 = 6,014 ÷ 8 =

6. 1,623 ÷ 9 = 3,508 ÷ 8 = 1,846 ÷ 6 = 3,522 ÷ 4 =

7. 18,834 ÷ 8 = 15,653 ÷ 4 = 15,680 ÷ 3 = 43,134 ÷ 5 =

8. 58,810 ÷ 7 = 24,404 ÷ 6 = 74,163 ÷ 9 = 14,235 ÷ 2 =

9. $\frac{8,335}{4} =$ $\frac{20,299}{5} =$ $\frac{52,022}{8} =$ $\frac{12,503}{6} =$

10. $\frac{15,014}{3} =$ $\frac{21,755}{7} =$ $\frac{14,057}{2} =$ $\frac{19,203}{4} =$

11. $\frac{54,725}{6} =$ $\frac{54,727}{9} =$ $\frac{40,053}{8} =$ $\frac{21,308}{3} =$

12. $\frac{5,530}{9} =$ $\frac{68,423}{7} =$ $\frac{83,212}{3} =$ $\frac{11,452}{6} =$

13. 21,815 / 8 = 98,167 / 5 = 62,189 / 4 = 44,981 / 7 =

14. 78,424 / 9 = 12,893 / 8 = 61,254 / 7 = 83,481 / 6 =

15. 12,290 / 3 = 19,538 / 5 = 38,242 / 4 = 22,517 / 8 =

Properties of Numbers

Before you tackle dividing by two-digit numbers, take the time to review all the basic operations with zeros and ones.

Do each of the following in your head.

1. $8 \div 1 =$	**2.** $1 + 6 =$	**3.** $5 - 1 =$	**4.** $9 + 0 =$
5. $7 \times 1 =$	**6.** $4 + 0 =$	**7.** $8 - 0 =$	**8.** $9 \times 1 =$
9. $0 \div 7 =$	**10.** $0 \times 4 =$	**11.** $1 - 0 =$	**12.** $5 \div 1 =$
13. $3 \times 0 =$	**14.** $6 + 1 =$	**15.** $1 \div 1 =$	**16.** $1 \times 7 =$
17. $0 + 8 =$	**18.** $1 \times 0 =$	**19.** $9 - 1 =$	**20.** $0 \div 3 =$

Be sure your answers are correct. Then read how these problems illustrate several properties about the basic operations.

For example, look at problems 2 and 14: $1 + 6 = 7$ and $6 + 1 = 7$. These two problems are examples of the **commutative property of addition.** In simple terms, the numbers in an addition problem can be added in any order.

Look at problems 5 and 16: $7 \times 1 = 7$ and $1 \times 7 = 7$. These two problems are examples of the **commutative property of multiplication.** The numbers in a multiplication problem can be multiplied in any order.

However, subtraction and division are *not* commutative. In problem 3, $5 - 1 = 4$, but $1 - 5$ is an algebra problem involving negative numbers. In problem 1, $8 \div 1 = 8$, but $1 \div 8$ is a fraction or decimal problem.

Look at problems 5, 8, and 16. These problems illustrate the fact that **any number multiplied by one is that number.**

Look at problems 1, 12, and 15. These problems illustrate the fact that **any number divided by one is that number.**

Look at problems 10, 13, and 18. These problems illustrate the fact that **any number multiplied by zero is zero.**

Look at problems 9 and 20. These problems illustrate the fact that **any number divided into zero is zero.** Notice that there is no problem such as $8 \div 0$. There is no number to multiply by zero to get 8. You cannot divide a number by zero. Mathematicians say that $8 \div 0$ is **undefined.**

Dividing by Two-Digit Numbers

Dividing by two-digit and three-digit numbers is a tricky process. It requires practice and a skill called **estimating;** that is, guessing how many times one number goes into another. Look at these examples carefully.

EXAMPLE 1

```
      46
32)1,472
   1 28
     192
     192
```

STEP 1 Ask yourself how many times 32 goes into 147. To estimate, ask how many times 30 goes into 147. $147 \div 30 = 4$ with a remainder.

STEP 2 Place the 4 over the 7 and multiply $4 \times 32 = 128$.

STEP 3 Subtract $147 - 128 = 19$.

STEP 4 Bring down the 2.

STEP 5 Ask yourself how many times 32 goes into 192. To estimate, ask how many times 30 goes into 192. $192 \div 30 = 6$ with a remainder.

STEP 6 Place the 6 over the 2 and multiply $6 \times 32 = 192$.

STEP 7 Subtract $192 - 192 = 0$.

CHECK

```
   46
 × 32
   92
 1 38
1,472
```

STEP 8 Check $32 \times 46 = 1{,}472$.

EXAMPLE 2

```
      58
47)2,726
   2 35
     376
     376
```

STEP 1 Ask yourself how many times 47 goes into 272. To estimate, ask how many times 50 goes into 272. $272 \div 50 = 5$ with a remainder.

STEP 2 Place the 5 over the 2 and multiply $5 \times 47 = 235$.

STEP 3 Subtract $272 - 235 = 37$.

STEP 4 Bring down the 6.

STEP 5 Ask yourself how many times 47 goes into 376. To estimate, ask how many times 50 goes into 376. $376 \div 50 = 7$ with a remainder.

STEP 6 Place the 7 over the 6 and multiply $7 \times 47 = 329$.

STEP 7 Subtract $376 - 329 = 47$.
Since 47 is left, 47 divides into 376 more than 7 times.
Erase the 7 and try 8. Multiply $8 \times 47 = 376$.

STEP 8 Subtract $376 - 376 = 0$.

CHECK

$$\begin{array}{r} 58 \\ \times\ 47 \\ \hline 406 \\ 2\,32 \\ \hline 2{,}726 \end{array}$$

STEP 9 Check $47 \times 58 = 2{,}726$.

Divide and check.

1. $42\overline{)336}$	$23\overline{)161}$	$62\overline{)310}$	$83\overline{)332}$
2. $52\overline{)312}$	$63\overline{)504}$	$72\overline{)288}$	$92\overline{)828}$
3. $19\overline{)133}$	$48\overline{)288}$	$57\overline{)228}$	$69\overline{)207}$
4. $87\overline{)174}$	$79\overline{)395}$	$38\overline{)228}$	$47\overline{)329}$
5. $22\overline{)176}$	$56\overline{)168}$	$63\overline{)252}$	$75\overline{)450}$

Rewrite each problem. Then divide and check.

6. $465 \div 93 =$ $432 \div 48 =$ $162 \div 54 =$ $460 \div 46 =$

7. $217 \div 31 =$ $108 \div 18 =$ $68 \div 17 =$ $672 \div 96 =$

8. $200 \div 25 =$ $320 \div 64 =$ $342 \div 38 =$ $318 \div 53 =$

9. $683 \div 82 =$ $283 \div 43 =$ $266 \div 65 =$ $132 \div 39 =$

10. $\frac{136}{29} =$ $\frac{66}{18} =$ $\frac{275}{43} =$ $\frac{207}{27} =$

11. $\frac{261}{46} =$ $\frac{353}{87} =$ $\frac{206}{52} =$ $\frac{598}{74} =$

12. $\frac{221}{66} =$ $\frac{191}{35} =$ $\frac{246}{56} =$ $\frac{201}{21} =$

13. $\frac{736}{32} =$ $\frac{1,845}{41} =$ $\frac{3,286}{53} =$ $\frac{5,022}{62} =$

14. 3,496 / 38 = 3,087 / 49 = 2,394 / 57 = 4,828 / 68 =

15. 2,352 / 98 = 2,736 / 76 = 667 / 29 = 3,696 / 88 =

16. $\frac{768}{19} =$ $\frac{1,687}{56} =$ $\frac{1,859}{37} =$ $\frac{1,742}{29} =$

17. $\frac{1,048}{20} =$ $\frac{1,650}{40} =$ $\frac{4,252}{80} =$ $\frac{567}{50} =$

18. 12,168 ÷ 52 = 13,546 ÷ 26 = 46,209 ÷ 73 = 20,482 ÷ 49 =

19. 20,368 ÷ 38 = 14,814 ÷ 18 = 25,480 ÷ 56 = 40,256 ÷ 64 =

20. $\frac{29,376}{72} =$ $\frac{44,616}{88} =$ $\frac{12,669}{41} =$ $\frac{32,224}{53} =$

21. $\frac{53,460}{99} =$ $\frac{17,280}{54} =$ $\frac{21,280}{28} =$ $\frac{18,130}{37} =$

Dividing by Three-Digit Numbers

EXAMPLE

```
        43
632)27,176
    25 28
     1 896
     1 896
```

STEP 1 Ask yourself how many times 632 goes into 2,717. (You know it won't go into 271.) To estimate, ask how many times 600 goes into 2,717.
$2,717 \div 600 = 4$ with a remainder.

STEP 2 Place the 4 over the 7 and multiply $4 \times 632 = 2,528$.

STEP 3 Subtract $2,717 - 2,528 = 189$.

STEP 4 Bring down the 6.

STEP 5 Ask yourself how many times 632 goes into 1,896. To estimate, ask how many times 600 goes into 1,800. $1,800 \div 600 = 3$

STEP 6 Place the 3 over the 6 and multiply $3 \times 632 = 1,896$.

STEP 7 Subtract $1,896 - 1,896 = 0$.

CHECK

```
   632
 ×  43
 1 896
25 28
27,176
```

STEP 8 Check $632 \times 43 = 27,176$.

Divide and check.

1. $326\overline{)7,824}$ $\quad 418\overline{)15,884}$ $\quad 521\overline{)26,571}$ $\quad 607\overline{)38,241}$

2. $862\overline{)16,022}$ $\quad 793\overline{)17,867}$ $\quad 486\overline{)22,566}$ $\quad 972\overline{)33,086}$

Rewrite each problem. Then divide and check.

3. $\frac{28{,}504}{509} =$ $\frac{30{,}456}{423} =$ $\frac{54{,}531}{657} =$ $\frac{23{,}867}{823} =$

4. $\frac{34{,}704}{964} =$ $\frac{79{,}807}{877} =$ $\frac{18{,}792}{216} =$ $\frac{17{,}490}{330} =$

5. 16,739 ÷ 418 = 37,712 ÷ 628 = 63,183 ÷ 902 = 22,606 ÷ 451 =

6. 37,659 ÷ 523 = 12,244 ÷ 139 = 25,611 ÷ 711 = 16,987 ÷ 269 =

7. 16,692 / 321 = 37,400 / 425 = 41,406 / 618 = 24,381 / 903 =

Rounding and Estimating

Not all division problems are as difficult as those in the last exercise. Numbers that end in zeros are often easy to work with.

EXAMPLE 1 $\frac{3{,}600}{9} = 400$

STEP 1 Divide 36 ÷ 9 = 4.

STEP 2 Bring along the zeros from 3,600.

Divide each problem in your head.

1. 1,200 ÷ 4 = 1,600 ÷ 8 = 2,100 ÷ 7 =

2. 2,000 ÷ 5 = 1,400 ÷ 2 = 40,000 ÷ 4 =

3. 480 ÷ 6 = 3,300 ÷ 3 = 7,200 ÷ 9 =

4. 630 ÷ 7 = 15,000 ÷ 5 = 30,000 ÷ 6 =

When both the dividend and the divisor end in zeros, you can *cancel* the zeros one-for-one.

EXAMPLE 2 $\frac{7{,}20\not{0}}{8\not{0}} = 90$

STEP 1 Divide 72 ÷ 8 = 9.

STEP 2 Cancel the zeros one-for-one, and bring along the remaining zero.

Divide each problem in your head.

5. 240 ÷ 40 = 8,100 ÷ 90 = 45,000 ÷ 90 =

6. 18,000 ÷ 200 = 1,800 ÷ 30 = 200 ÷ 20 =

7. 2,400 ÷ 600 = 4,900 ÷ 70 = 3,500 ÷ 70 =

8. 15,000 ÷ 50 = 2,400 ÷ 300 = 64,000 ÷ 80 =

Sometimes you can round the dividend to a number that is easy to divide into. You can use the rounded number to estimate an answer to the original division problem.

EXAMPLE 3 Estimate the answer to the problem $539 \div 9$.

$539 \div 9 \approx 540 \div 9$	STEP 1	Round 539 to the nearest ten.
$539 \div 9 \approx 500 \div 9$	STEP 2	Round 539 to the nearest hundred.
$540 \div 9 = \mathbf{60}$	STEP 3	Since 540 divides evenly by 9, use the problem $540 \div 9$ to estimate the answer.

EXAMPLE 4 Estimate the answer to the problem $209 \div 4$.

$209 \div 4 \approx 210 \div 4$	STEP 1	Round 209 to the nearest ten.
$209 \div 4 \approx 200 \div 4$	STEP 2	Round 209 to the nearest hundred.
$200 \div 4 = \mathbf{50}$	STEP 3	Since 200 divides evenly by 4, use the problem $200 \div 4$ to estimate the answer.

Round each dividend to the nearest *ten* and the nearest *hundred*. Then decide which number is easier to divide. Use the easier rounded number to estimate the answer.

9. $423 \div 7 \approx$ $\qquad$ $409 \div 4 \approx$

10. $329 \div 6 \approx$ $\qquad$ $492 \div 7 \approx$

11. $194 \div 5 \approx$ $\qquad$ $958 \div 8 \approx$

12. $212 \div 3 \approx$ $\qquad$ $716 \div 9 \approx$

Round each dividend to the nearest *hundred* and the nearest *thousand*. Then decide which rounded number is easier to divide. Use the easier rounded number to estimate the answer.

13. $3{,}186 \div 4 \approx$ $\qquad$ $4{,}236 \div 8 \approx$

14. $3{,}186 \div 6 \approx$ $\qquad$ $2{,}095 \div 7 \approx$

15. $8{,}077 \div 9 \approx$ $\qquad$ $3{,}182 \div 5 \approx$

16. $1{,}790 \div 3 \approx$ $\qquad$ $4{,}773 \div 6 \approx$

Two-Digit Accuracy

Another way to estimate answers is to do a partial division. Instead of completing a division problem, try for **two-digit accuracy.** Divide until you have three digits in the quotient. Then round your answer. Study the example carefully.

EXAMPLE **Estimate an answer to the problem 158,916 ÷ 41.**

```
    3,87– ≈ 3,900
41)158,916
   123
    35 9
    32 8
     3 11
```

STEP 1 Ask yourself how many times 41 goes into 158. To estimate, ask how many times 40 goes into 158. $158 \div 40 = 3$ with a remainder.

STEP 2 Place 3 over the 8 and multiply $3 \times 41 = 123$.

STEP 3 Subtract $158 - 123 = 35$.

STEP 4 Bring down the 9.

STEP 5 Ask yourself how many times 41 goes into 359. To estimate, ask how many times 40 goes into 359. $359 \div 40 = 8$ with a remainder.

STEP 6 Place 8 over the 9 and multiply $8 \times 41 = 328$.

STEP 7 Subtract $359 - 328 = 31$.

STEP 8 Bring down the 1. Then decide whether 41 goes into 311 five times or more. $31 \div 4 = 7$ with a remainder. Since the digit to the right of 8 is 5 or more, raise 8 to 9, and put zeros in the tens and units places.

Calculate each problem to two-digit accuracy.

1. $\frac{7,936}{62} \approx$ $\frac{18,316}{76} \approx$ $\frac{14,798}{49} \approx$ $\frac{34,506}{81} \approx$

2. $\frac{20,909}{29} \approx$ $\frac{45,724}{92} \approx$ $\frac{18,444}{53} \approx$ $\frac{23,834}{34} \approx$

3. $\frac{105,705}{87} \approx$ $\frac{157,212}{66} \approx$ $\frac{144,648}{41} \approx$ $\frac{569,439}{93} \approx$

Use your knowledge of two-digit accuracy to select the correct answer from the choices.

4. 5,943 ÷ 21 =

a. 83
b. 183
c. 213
d. 283

5. 23,072 ÷ 32 =

a. 961
b. 721
c. 551
d. 341

6. 22,684 ÷ 53 =

a. 618
b. 538
c. 428
d. 298

7. 19,034 ÷ 62 =

a. 417
b. 397
c. 307
d. 267

8. 65,807 ÷ 79 =

a. 833
b. 763
c. 703
d. 683

9. 13,708 ÷ 46 =

a. 488
b. 438
c. 378
d. 298

10. 154,686 ÷ 58 =

a. 1,927
b. 2,667
c. 3,107
d. 3,577

11. 365,556 ÷ 82 =

a. 3,998
b. 4,108
c. 4,458
d. 5,018

12. 261,501 ÷ 67 =

a. 2,543
b. 2,933
c. 3,023
d. 3,903

13. 448,622 ÷ 82 =

a. 5,471
b. 5,991
c. 6,301
d. 6,781

14. 213,014 ÷ 73 =

a. 3,388
b. 2,918
c. 2,278
d. 1,858

15. 108,824 ÷ 61 =

a. 994
b. 1,254
c. 1,784
d. 2,164

Applying Your Division Skills

The problems in this exercise require you to apply your division skills. In each problem, pay close attention to the language that tells you to divide.

You may be asked how many of one thing are contained in something else.

EXAMPLE 1 **A truck can carry 2,000 pounds. How many boxes, each weighing 50 pounds, can the truck carry?**

$\frac{2{,}00\not{0}}{5\not{0}} =$ **40 boxes** Find how many times 50 divides into 2,000.

You may be given information about several things and asked for information about one of those things.

EXAMPLE 2 **Hector makes $84 for working 7 hours for a moving company. How much did he make each hour?**

$\frac{\$84}{7} =$ **$12** Divide the total wages, $84, by the number of hours he worked, 7.

You may be asked to find an average. An **average,** or **mean,** is a total number divided by the number of things that make up the total.

EXAMPLE 3 **Together, three children weigh 201 pounds. What is the average weight of each child?**

67 ⟵ average weight

number of children ⟶ 3)201 ⟵ total weight

Divide 201 by 3.

Solve and write the correct label, such as miles or pounds, next to each answer.

1. A dozen folding chairs cost $357. What is the price of one chair?

2. Estimate the price of one chair by rounding the total cost to the nearest *ten* dollars.

3. One season a professional basketball player scored 1,638 points. He played in 78 games that season. What average number of points did he score per game?

4. To estimate the player's average points per game, round the total number of points he scored to the nearest *hundred,* and round the number of games he played to the nearest *ten.*

5. The State Theater has 1,584 seats. If each row has 36 seats, how many rows are there in the theater?

6. Estimate the number of rows in the last problem by rounding the total number of seats to the nearest *hundred* and the number of seats per row to the nearest *ten.*

7. Last year the Dean family spent $477.60 for gas and electricity. What was the family's average monthly cost for gas and electricity?

8. Estimate the Dean family's monthly bill by first rounding the yearly total to the nearest *ten* dollars.

9. Three pounds of roast beef cost $17.97. Find the price of one pound of beef.

10. Estimate the price of one pound of roast beef in the last problem by rounding the price of three pounds to the nearest whole number of dollars.

11. The Sotomayor family spends $6,894 a year for rent. What is their monthly rent?

12. An airplane traveled at an average speed of 365 miles per hour. How many hours did the plane take to travel a distance of 5,110 miles?

13. At her part-time job at a day-care center, Rosa makes $22,672 in a year. How much does she make each week? (1 year = 52 weeks)

14. The Jackson family drove a total of 5,376 miles on their vacation. If they traveled for 14 days, what average distance did they drive each day?

15. A paint store owner has 1,148 quarts of red paint. To save space, he wants to transfer the paint to one-gallon containers. There are 4 quarts in 1 gallon. How many one-gallon containers will he need?

16. On a regional map, one inch is equal to 35 miles. How many inches apart on the map are two cities that are actually 455 miles apart?

17. Nam worked four months last year for a tree-trimming service where he made $9,583. To the nearest *hundred* dollars, what was Nam's average monthly income at the tree-trimming service?

18. After completing a speed reading class, you can now read 285 words per minute. Which of the following is closest to the number of minutes you will need to read an article that contains 12,000 words?

a. 20 minutes
b. 30 minutes
c. 40 minutes
d. 50 minutes

19. The total rainfall one year in San Juan, Puerto Rico, was 60 inches. What was the average monthly rainfall?

20. Serena paid a total of $1,176 for a new laptop computer. If she paid monthly installments of $98, how many months did it take her to pay for the computer?

21. The Saba family and two other families in their neighborhood agree to share 24 cubic yards of topsoil for their gardens. If they share equally, how many cubic yards of topsoil will each family receive?

22. A gallon of the paint Marlene is using to paint two rooms in her house will cover 250 square feet. She estimates that the area of the walls and ceilings she wants to paint is about 1,500 square feet. How many gallons of paint will she need to complete her job?

23. Alex drove his van 435 miles in 7 hours. To the nearest *ten,* what was his average speed in miles per hour?

24. Tim drove his compact car 12,664 miles last year. He gets about 25 miles per gallon of gasoline. To the nearest *ten,* how many gallons of gasoline did Tim have to buy last year?

The answers to most division problems have remainders, but sometimes you can ignore the remainders. Think about the next two examples carefully.

Kaliska works in a shipping department where she packs ballpoint pens. Each box contains 6 pens. How many boxes will she need to ship an order of 100 pens?

$100 \div 6 = 16$ with a remainder

Divide the total order, 100, by the number of pens in each box, 6. Kaliska can fill 16 boxes, but she needs **17 boxes** to ship the order.

EXAMPLE 5 **How many CDs at the price of $14 each can Hendri buy with $100?**

$100 \div 14 = 7$ with a remainder

Divide Hendri's money, $100, by the cost of one CD, $14.
Hendri will have a little money left, but he can buy only **7 CDs.**

25. A salesperson told Audrey that she can pay $48 a month for the furniture she wants. How many months will she need to pay for furniture that costs $1,219?

26. Each shelf in Fernando's grocery store holds 24 cans of vegetables. How many shelves does Fernando need to display a shipment of 500 cans?

27. A freight elevator will safely hold 4,000 pounds. If one crate weighs 150 pounds, how many crates can the elevator hold?

28. Inez is a packer in the shipping department of a book publisher. She can put 36 copies of a new book in one carton. How many cartons does she need to ship an order of 1,000 books?

29. Mark borrowed $500 from his brother David. Mark agreed to return the money to his brother in $60 monthly payments. How many months will Mark need to repay his brother?

30. Manny bought a used car. If he makes payments of $85 a month, how many months will it take him to pay off a car loan of $2,750?

Multistep Problems

The problems in this exercise require that you use different combinations of the four basic operations you have studied in this book. Read these problems carefully to be sure that you answer the question that is asked.

Calculating an **average,** or **mean,** is a common multistep application using addition and division. An average is a number that is typical of a group of numbers.

To find the average of a group of numbers, add the numbers together. Then divide the total by the number of numbers in the group.

EXAMPLE On her Spanish quizzes, Sue got scores of 80, 85, and 93. Find the average.

STEP 1 Add the scores. $80 + 85 + 93 = 258$

STEP 2 Divide the total by the number of scores, 3. $258 \div 3 = 86$

Sue's average score on the three quizzes was **86.**

Solve and write the correct label, such as $ or miles, next to each answer.

1. One month, a waiter worked 40 hours the first week, 25 hours the second week, 45 hours the third week, and 30 hours the fourth week. Find the average number of hours he worked each week.

2. In a year, Blanca makes a gross salary of $42,785 as a supervisor in a bank. Her employer deducts $8,556 for taxes and social security from her gross pay. To the nearest *hundred* dollars, what is her monthly take-home pay?

3. Pete drives a cab five evenings a week. On Monday he drove 3 hours; on Tuesday, 6 hours; on Wednesday, 5 hours; on Thursday, 4 hours; and on Friday, 7 hours. What average number of hours did he drive each evening?

4. On Monday evening, Pete made $24.30 in tips; on Tuesday, $68.55; on Wednesday, $56.15; on Thursday, $49.65; and on Friday, $92.35. How much did Pete average in tips each evening?

5. Miguel and Ann pay $780 a month for their two-bedroom apartment. Miguel is a full-time student, and Ann takes home $31,686 a year as a nurse. Calculate how much they have left each month after they pay rent.

6. Elizabeth received the following scores on math tests: 86, 76, 93, 89, 68, and 92. What was the average of her math test scores?

7. Todd commutes to work by train. He makes 20 round-trip rides every month, and he pays $640 for a monthly pass. A one-way ticket costs $19. How much does Todd save on each ride by using his monthly pass?

8. The tenants' association at Kendal Houses meets every Tuesday night. On the 4th, there were 46 people at the meeting. On the 11th, there were 53; on the 18th, 38; and on the 25th, 55. What was the average attendance for the meetings that month?

9. The Perez family took three days to drive to their grandparents' house in the Midwest. Thursday they drove 487 miles, Friday they drove 392 miles, and on Saturday they drove 456 miles. Find the average distance they drove each day.

10. Jim and Maya want to buy a condominium that is selling for $129,500. They plan to make a down payment of $30,000. Calculate, to the nearest *ten,* the number of months it will take them to pay off the balance that they will owe if they make monthly payments of no more than $700.

11. Five clerk typists, seven file clerks, and three receptionists pooled their resources to buy lottery tickets. They won $100,000. They will share the winnings equally. To the nearest *hundred* dollars, how much will each person get?

12. In 2006 Mr. Lee made $36,776. In 2007 he made $39,231. In 2008 he made $28,385, and in 2009 he made $37,595. Round each year's income to the nearest *hundred* dollars. Then estimate Mr. Lee's average annual income for those years.

13. In 2006 Mr. Lee's family consisted of his wife, their two children, and himself. Using the exact income, find the average (per capita) income for each member of the Lee family in 2006.

14. In 2009 the Lees had another child. Using the exact income, find the per capita income for each member of the Lee family in 2009.

Division Review

This review covers the material you have studied so far in this book. Read the signs carefully to decide whether to add, subtract, multiply, or divide. When you finish, check your answers at the back of the book

1. The factory where Casey works produced 13,826 machine parts in one month. Round the number of parts to the nearest *thousand*. 14,000 ✓

2. What is the value of the digit 8 in the number 13,826? 800 ✓

3. 236 + 4,807 = 5,043 ✓

4. 26,458 + 9,583 = 36,041 ✓

5. 16 + 56 + 384 = 456 ✓

6. 8,134 − 2,023 = 6,111 ✓

7. 25,126 − 19,587 = 5,539 ✓

8. 250,000 − 83,056 = 166,944 ✓

9. 6 × 53 = 318 ✓

10. (24)(87) = 2,088 ✓

11. 36 · 258 = 9,288 ✓

12. (724)(1,000) = 724,000 ✓

13. 258 ÷ 6 = 43 ✓

14. 5,022 / 8 = 627 R6 ✓

15. 496 ÷ 62 = 8 ✓

16. $\frac{3,096}{43}$ = 72 ✓ ok

17. 33,369 ÷ 49 = 681 ✓

18. 20,808 ÷ 289 = 72 ✓

19. Round both numbers to the nearest *hundred* and multiply.

$(293)(719) \approx$

20. Round the larger number to the nearest *thousand*. Then divide by 9.

$71{,}648 \div 9 \approx$

21. A plane flew at an average speed of 415 miles per hour for 7 hours. How far did the plane travel?

22. Once a month a maintenance crew has to wax the floors of the conference rooms in a large office. The conference rooms have a combined floor area of 7,800 square feet. Each can of wax will cover 600 square feet. How many cans of wax does the crew need to do the floors of the conference rooms?

23. Margaret earned $40,260 last year. To the nearest *ten* dollars, how much did she earn each week? (52 weeks = 1 year)

24. Judy drove 989 miles on a trip to North Carolina. She had to buy 43 gallons of gasoline on the trip. How many miles per gallon did she average on the trip?

25. During his three-month vacation from school, Alberto did yard work. In June he made $1,580; in July, $2,910; and in August, $1,876. To the nearest *hundred* dollars, what was his average monthly income for the summer?

DIVISION REVIEW CHART

If you missed more than one problem on any group below, review the practice pages for those problems. Then redo the problems you got wrong before going on to Posttest A. If you had a passing score, redo any problem you missed and begin Posttest A on page 114.

Problem Numbers	Skill Area	Practice Pages
1, 2	place value	6–12
3, 4, 5	addition	15–28
6, 7, 8	subtraction	33–50
9, 10, 11, 12, 21	multiplication	56–78
13, 14	dividing by 1 digit	87–89
15, 16, 17, 18	dividing by 2 and 3 digits	94–99
19, 20	rounding and estimating	100–101
22, 23, 24, 25	applying division	104–110

Posttest A

This posttest gives you a chance to check your basic mathematical skills. Take your time and work each problem carefully. When you finish, check your answers and review any topics on which you need more work.

1. Which digit in the number 56,324 is in the thousands place?

2. What is the value of the digit 3 in 56,324?

3. The population of Midvale is 287,037. Round the number to the nearest *ten thousand.*

4. In a recent year, the state of Texas spent \$3,692 per student in their public school system. Round \$3,692 to the nearest *hundred* dollars.

5. $\begin{array}{r} 126 \\ +\ 543 \\ \hline \end{array}$

6. $\begin{array}{r} 12{,}082 \\ 9{,}565 \\ +\quad 788 \\ \hline \end{array}$

7. $9 + 84 + 71 + 6 =$

8. $125{,}036 + 84{,}591 =$

9. $76{,}053 + 248 + 1{,}592 =$

10. Round 2,724 to the nearest *thousand.* Round 809 to the nearest *hundred.* Round 19,628 to the nearest *ten thousand.* Then add.

11. The driving distance from Atlanta to Washington, D.C., is 630 miles. The distance from Washington, D.C., to New York City is 229 miles. The distance from New York City to Boston is 216 miles. Find the total driving distance from Atlanta to Boston by way of Washington, D.C., and New York.

12. Maria and Tom were recently married. From her savings, Maria will spend $11,200 as a down payment for a new house. Tom can contribute $7,500, and Tom's parents promise to give the couple $12,000. Find the total amount Maria and Tom have for their down payment.

13. $\begin{array}{r} 358 \\ -\,216 \\ \hline \end{array}$

14. $\begin{array}{r} 8{,}542 \\ -\,5{,}296 \\ \hline \end{array}$

15. 2,643 − 1,978 =

16. 300,000 − 78,140 =

17. 460,200 − 381,530 =

18. Round each number to the nearest *thousand* and subtract.

49,528 − 17,485 ≈

19. The price of a popular compact car ranges from $14,250 for the basic model to $22,819 for the model with all the accessories. The basic model is how much less than the model with all the accessories?

20. Clarkville Central School hopes to raise $100,000 for scholarships. So far they have raised $83,965. How much more does the school need to raise?

21. $\begin{array}{r} 23 \\ \times\ 32 \\ \hline \end{array}$

22. $\begin{array}{r} 1{,}984 \\ \times\ \quad 7 \\ \hline \end{array}$

23. $89 \times 67 =$

24. $1{,}000 \cdot 382 =$

25. $(1{,}736)(73) =$

26. Round 48 to the nearest *ten* and 3,280 to the nearest *thousand.* Then multiply.

27. On a state map 1 inch represents a distance of 35 miles. Two towns are 5 inches apart on the map. What is the actual distance in miles between the two towns?

28. The owner of a car dealership must pay a supplier $196 for a popular model of an AM/FM/CD player. How much will he pay the supplier for an order of 15 of the players?

29. $6\overline{)468}$

30. $8\overline{)7,237}$

31. $4,067 \div 49 =$

32. $\frac{41,400}{90} =$

33. $16,132 / 218 =$

34. Find the answer to the problem $395,884 \div 76$ to the nearest *thousand.*

35. An auditorium has 1,610 seats. Every row in the auditorium has 35 seats. How many rows of seats are there in the auditorium?

36. A warehouse received a shipment that weighed a total of 3,723 pounds. The shipment consisted of identical boxes of small motor parts, each weighing 17 pounds. How many boxes were in the shipment?

POSTTEST A CHART

Circle the number of any problem that you miss. If you missed one or less in each group below, go on to Using Number Power on page 119. If you missed more than one problem in any group, review the practice pages for those problems. Then redo any problems you missed.

Problem Numbers	Skill Area	Practice Pages
1, 2, 3, 4	place value	6–12
5, 6, 7, 8, 9, 10, 11, 12	addition	15–28
13, 14, 15, 16, 17, 18, 19, 20	subtraction	33–50
21, 22, 23, 24, 25, 26, 27, 28	multiplication	56–78
29, 30, 31, 32, 33, 34, 35, 36	division	84–110

USING NUMBER POWER

Changing Units of Measurement

This page and the following four pages will give you practice working with units of measurement. To change from one unit of measurement to another, refer to the table on page 172. It is not necessary to memorize the table, but you should become familiar with the measures listed there.

To change **from a large unit** of measurement **to a small unit, multiply** by the number of small units contained in one large unit.

EXAMPLE 1 **Change 6 pounds to ounces.**

STEP 1 Check the table of measurements on page 172 to find out how many of the small units are contained in one large unit. 1 pound = 16 ounces

STEP 2 Multiply $16 \times 6 = 96$ ounces.

To change **from a small unit** of measurement **to a large unit, divide** by the number of small units contained in one large unit.

EXAMPLE 2 **Change 48 inches to feet.**

STEP 1 Check the table of measurements on page 172 to find out how many of the small units are contained in one large unit. 1 foot = 12 inches

STEP 2 Divide. $12\overline{)48}$ = 4

Change each of the following to the unit indicated. Use the table of measurements on page 172 to find out how many small units are contained in the large units in each problem.

1. 192 ounces = ________ pounds

2. 5 miles = ________ feet

3. 4,000 meters = ________ kilometers

4. 54 feet = ________ yards

5. 72 gallons = ________ quarts

6. 12 kilograms = ________ grams

7. 504 inches = ________ yards

8. 8 hours = ________ minutes

9. 425 meters = ________ centimeters

10. 296 cups = ________ quarts

11. 15,840 yards = ________ miles

12. 420 quarts = ________ gallons

Use the tables on page 172 to solve each problem.

13. The driveway at Mr. and Mrs. Cho's house is 120 feet long. What is the length of the driveway in yards?

14. Janina is preparing soup for a neighborhood dinner. How many gallon buckets can she fill with 24 quarts of soup?

15. Fred's desk is two meters long. What is the length of the desk in centimeters?

16. Jean ordered 12 cubic yards of topsoil for a new vegetable garden. Find the volume of the topsoil in cubic feet.

17. Minh has to install wall-to-wall carpet in a room with a floor area that measures 180 square feet. Find the area of the floor in square yards.

18. A package weighs 48 ounces. What is the weight of the package in pounds?

19. Tony's pickup truck can carry a maximum weight of three tons. What total number of pounds can the truck carry?

20. A fish tank has a volume of 924 cubic inches. How many gallons of water does it hold?

21. Antonio is exactly six feet tall. What is his height in inches?

22. When he dug a hole for a new concrete patio, Carl removed 135 cubic feet of soil. How many cubic yards of soil did he remove?

23. How many square feet is the surface of a table that has an area of 2,592 square inches?

Adding Measurements

EXAMPLE

2 ft 9 in. 3 ft 10 in. + 1 ft 8 in. 6 ft 27 in.	STEP 1	Add each unit of measurement separately.
2 ft 3 in. 12)27 24 3	STEP 2	If you have enough of a small unit to change it to a larger unit, change it by dividing. Use the Table of Measurements to help you.
6 ft + 2 ft 3 in. 8 ft 3 in.	STEP 3	Combine your results.

Add the following measurements.

1. 2 hr 38 min
+ 6 hr 47 min

2. 2 gal 3 qt
+ 5 gal 3 qt

3. 4 lb 12 oz
+ 7 lb 11 oz

4. 8 yd 2 ft
+ 6 yd 1 ft

5. 3 days 18 hr
+ 8 days 16 hr

6. 2 m 75 cm
+ 3 m 45 cm

7. 8 ft 4 in.
3 ft 11 in.
+ 2 ft 9 in.

8. 4 kg 800 g
6 kg 250 g
+ 3 kg 415 g

9. 10 min 15 sec
5 min 52 sec
+ 9 min 47 sec

10. 3 T 1,500 lb
4 T 850 lb
+ 5 T 1,175 lb

11. 5 wk 4 days
6 wk 6 days
+ 4 wk 5 days

12. 1 day 10 hr
3 days 15 hr
+ 4 days 12 hr

Subtracting Measurements

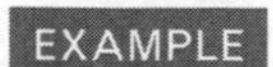

EXAMPLE

$$\begin{array}{r} 7 \text{ hr } 35 \text{ min} \\ -\ 4 \text{ hr } 50 \text{ min} \\ \hline \end{array}$$

STEP 1 You can't take a larger number from a smaller number, so you must borrow. Use the Table of Measurements to find out how many smaller units there are in the larger unit from which you must borrow. Borrow one larger unit changed to smaller units (1 hr = 60 min) and add it to the smaller units you already have (60 + 35 = 95 min).

$$\begin{array}{r} {}^{6}\not{7} \text{ hr } \overset{95}{\not{35}} \text{ min} \\ -\ 4 \text{ hr } 50 \text{ min} \\ \hline 2 \text{ hr } 45 \text{ min} \end{array}$$

STEP 2 Subtract each column.

Subtract the following measurements.

1. 8 ft 5 in. − 4 ft 8 in.
2. 5 days 11 hr − 3 days 18 hr
3. 13 lb 8 oz − 10 lb 14 oz
4. 12 gal 1 qt − 9 gal 3 qt
5. 9 m 45 cm − 5 m 92 cm
6. 8 min 25 sec − 2 min 53 sec
7. 8 hr 15 min − 5 hr 50 min
8. 3 T 800 lb − 2 T 1,560 lb
9. 6 km 375 m − 3 km 680 m
10. 5 yd − 2 yd 2 ft
11. 9 m − 8 m 36 cm
12. 9 wk − 4 wk 2 days

Multiplying Measurements

EXAMPLE

```
  4 lb 10 oz
×          3
 12 lb 30 oz
```

STEP 1 Multiply each unit of measurement separately.

```
      1 lb 14 oz
16)30
   16
   14
```

STEP 2 If you have enough of a small unit to change it to a larger unit, change it by dividing. Use the Table of Measurements to help you.

```
  12 lb
+  1 lb 14 oz
  13 lb 14 oz
```

STEP 3 Combine your results.

Multiply the following measurements.

1. 2 ft 8 in. × 4

2. 2 gal 5 qt × 6

3. 8 lb 12 oz × 5

4. 6 T 350 lb × 10

5. 8 m 36 cm × 4

6. 4 wk 3 days × 6

7. 2 yd 8 in. × 9

8. 3 km 600 m × 12

9. 7 hr 25 min × 3

10. 4 kg 640 g × 8

11. 18 yd 2 ft × 7

12. 5 days 18 hr × 6

Dividing Measurements

EXAMPLE

$$\begin{array}{r|l} & \text{1 day} \quad \text{13 hours} \\ 4) & \text{6 days} \quad \text{4 hours} \\ & \text{4 days} \\ & \text{2 days} = \text{48 hours} \\ & +4 \\ & \text{52 hours} \\ & 52 \\ & 0 \end{array}$$

STEP 1 Divide into the first unit.
$6 \div 4 = 1$ with a remainder of 2.

STEP 2 Change the remainder into the next type of units.

STEP 3 Add to the units you already have.

STEP 4 Divide again $52 \div 4 = 13$.

Divide the following measurements.

1. $6\overline{)8 \text{ lb } 4 \text{ oz}}$

2. $4\overline{)9 \text{ yd } 1 \text{ ft}}$

3. $5\overline{)17 \text{ gal } 2 \text{ qt}}$

4. $9\overline{)12 \text{ T } 300 \text{ lb}}$

5. $6\overline{)13 \text{ km } 800 \text{ m}}$

6. $4\overline{)15 \text{ ft } 8 \text{ in.}}$

7. $5\overline{)19 \text{ lb } 6 \text{ oz}}$

8. $12\overline{)30 \text{ wk } 6 \text{ days}}$

9. $13\overline{)28 \text{ m } 34 \text{ cm}}$

10. $17\overline{)36 \text{ min } 16 \text{ sec}}$

11. $4\overline{)9 \text{ kg } 420 \text{ g}}$

12. $11\overline{)24 \text{ hr } 45 \text{ min}}$

Perimeter: Measuring the Distance Around a Rectangle

Most doors, windows, floors, and pieces of paper are in the shape of **rectangles.** A rectangle is a flat figure with four sides. The sides across from each other are equal in length and are also **parallel,** which means that they remain the same distance from each other. The long side of a rectangle is called the **length,** and the short side is called the **width.**

The distance around a rectangle is called the **perimeter.** To calculate the perimeter of a rectangle, add all four sides together.

EXAMPLE 1 **Find the perimeter of the figure below.**

length = 22 in.

width = 10 in.

STEP 1 Find out how long each side is. Since the sides across from each other are equal in length, you have two 22-inch sides and two 10-inch sides.

STEP 2 Add the lengths of all four sides.
$22 + 22 + 10 + 10 =$ 64-inch perimeter

Another way to find the distance around a rectangle is to double the length and double the width. Then add these two numbers together.

For the example above

2×10 in. $= 20$ in.
2×22 in. $= 44$ in.
20 in. $+ 44$ in. $=$ 64-inch perimeter

EXAMPLE 2 **The distance around the outside of a room measures 52 feet. The width is 9 feet. How long is the room?**

perimeter = 52 ft

width = 9 ft

In this problem, you already know the perimeter. This means that you know the total distance around the rectangle. You also know the measurement of two of the four sides. You need to find the measurement of one of the two remaining sides.

STEP 1 The distance covered by the two widths is $9 \text{ ft} + 9 \text{ ft} = 18 \text{ ft}$.

STEP 2 The distance covered by the two lengths is $52 \text{ ft} - 18 \text{ ft} = 34 \text{ ft}$.

STEP 3 The distance covered by one length is $34 \text{ ft} \div 2 = 17 \text{ ft}$.

Solve and write the correct label, such as $ or feet, next to each answer.

1. How much fencing is required to enclose a garden that is 18 feet wide and 23 feet long?

2. Welded wire garden fencing costs $1.29 per foot. Find the cost of the fencing for the garden in the problem above.

3. Mr. Fedea wants to put weather stripping around the large windows of his house. Each window is 3 feet wide and 5 feet high. There are 11 of these windows in his house. How many feet of weather stripping must Mr. Fedea buy?

4. Foam weather strip tape costs 39¢ per foot. How much will it cost to buy weather strip tape for the windows of Mr. Fedea's house?

5. Mr. Ellis wants to fence in part of his backyard for his young children. He has 80 feet of fencing to use. If the rectangular space he encloses is 17 feet wide, how long can it be?

6. Tim works in an art supply store as a frame maker. How many inches of molding are needed to make a frame for a painting that is 24 inches wide and 36 inches long?

7. What is the total length in feet of the molding for the frame in problem 6?

8. Mr. Romney's rectangular garden required 82 feet of fencing to completely enclose it. If the garden is 27 feet long, how wide is it?

Area: Measuring the Space Inside a Rectangle

When you want to decide how much carpeting to buy to cover your floor, or how much paint you need to cover a wall, you need to find **area.** Area is the measure of the amount of space inside a flat figure.

Area is measured in **square units:** square inches, square feet, square yards, square miles, and so on.

To find the area of a rectangle, multiply the length by the width.

EXAMPLE 1 Find the area of the rectangle shown below.

length = 5 in.

width = 3 in.

Area = 5 in. × 3 in. = 15 square inches

By dividing the rectangle in Example 1 into one-inch squares, you can see that there are 15 one-inch squares inside the rectangle.

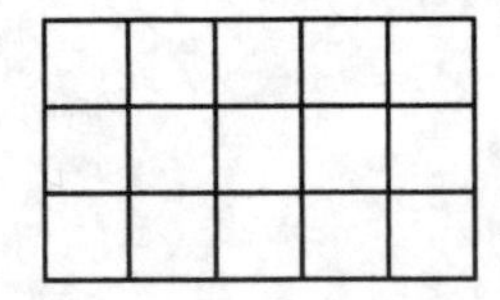

To find the measurement of one side of a rectangle when the area and the measurement of the other side are given, *divide* the area by the side of the rectangle you know.

EXAMPLE 2 A room requires 240 square feet of carpeting to cover the floor. If you know that the room is 20 feet long, how wide is it?

length = 20 ft

width = ?

Area = 240 sq ft

```
    12 feet wide
20)240
   20
    40
    40
```

Solve and write the correct label, such as $ or square feet, next to each answer.

1. What is the area of the floor of a room that is 12 feet wide and 18 feet long?

2. Find the area in square yards for the room in problem 1.
(1 square yard = 9 square feet)

3. If a carpet costs $16.99 per square yard, how much would it cost to buy carpet for the room in problem 1?

4. A store charges $75 to deliver and install carpet. Find the total cost including delivery and installation for the carpet in the last problem.

5. Mr. Cortez installs vinyl flooring. He was hired to put vinyl tiles on a basement floor. Each tile covers 1 square foot. The basement floor is 63 feet long and 24 feet wide. How many tiles does he need to cover the floor?

6. If the tiles Mr. Cortez plans to use cost $0.78 each, how much will the tiles for the basement floor cost?

7. Mr. Cortez needs 4 gallons of adhesive at $29.95 per gallon to attach the tiles. Find the combined cost of the tiles and the adhesive.

8. A group of neighbors plans to paint a wall that surrounds part of the playground of their community center. One gallon of paint is supposed to cover 300 square feet. If the wall is 12 feet high, what is the length of a section of the wall that can be covered by 1 gallon of paint?

9. The wall that the group wants to paint is 175 feet long. How many gallons of paint are needed to cover the wall?

10. The paint for the wall in the last two problems costs $19.95 per gallon. Find the total cost of the paint needed to cover the wall.

11. The area of the gym floor inside a community center is 7,168 square feet. If the gym floor is 64 feet wide, how long is it?

12. Find the area of the fenced-in play area in Mr. Ellis's yard in problem 5 on page 127.

13. Find the area of the picture in problem 6 on page 127.

14. What is the area of Mr. Romney's garden in problem 8 on page 127?

Volume: Measuring the Space Inside a 3-Dimensional Object

Shapes like boxes, suitcases, and rooms have three dimensions. Besides length and width, these shapes also have height or depth. A measure of the amount of space inside a three-dimensional object is called **volume.**

Volume is measured in **cubic units:** cubic inches, cubic feet, cubic yards, and so on.

To find the volume of a 3-dimensional object, such as a box or a trunk, multiply the length by the width by the height.

EXAMPLE

length = 20 in.
width = 8 in.
height = 7 in.

STEP 1 Multiply the length by the width.

$$\begin{array}{r} 20 \\ \times\ 8 \\ \hline 160 \end{array}$$

STEP 2 Multiply this number by the height.

$$\begin{array}{r} 160 \\ \times\ 7 \\ \hline 1{,}120 \end{array}$$ 1,120 cubic inches (cu in.)

Solve and write the correct label, such as $ or cubic inches, next to each answer.

1. Mr. Miller is a builder. He is adding a room to the back of a house. For the foundation, he must dig a hole that is 16 feet long, 12 feet wide, and 4 feet deep. How many cubic feet of soil have to be removed?

2. Find the volume of a shoe box that is 9 inches wide, 15 inches long, and 6 inches high.

3. Mrs. Rodriguez's ice tray is 1 inch deep, 9 inches long, and 3 inches wide. How many cubic inches of ice will the tray hold?

4. Mrs. Rodriguez's tray makes ice cubes that are each 1 cubic inch. How many cubes can be made with five trays?

5. When the Corona family goes on vacation, they always pack one big trunk instead of several small suitcases. Their trunk measures 6 feet long by 4 feet wide by 3 feet high. Find the volume of the packing space of the trunk.

6. The storeroom of the Apex Company is 30 feet long and 24 feet wide and has a 12-foot high ceiling. How many cubic feet of storage space does the room contain?

7. How many cartons each measuring 3 feet by 3 feet by 2 feet can fit in the Apex Company storeroom?

8. To fill a hole in a vacant lot that is 12 feet long, 10 feet wide, and 10 feet deep, the community council hired a trucking company that could carry 40 cubic feet of dirt per truckload. How many truckloads will it take to fill the hole?

9. The trucking company charges $25 for the delivery of each truckload of dirt. What will be the cost of filling the hole in the last problem?

10. The book storage room in a training center is 16 feet long, 12 feet wide, and 8 feet high. How many boxes each measuring 4 feet by 4 feet by 2 feet can be put in the storage room?

Pricing a Meal from a Menu

Use the menu below to answer the questions on this page.

Lois's Restaurant Menu

Soup of the day	$3.85	Coffee	$1.35
Tossed salad	$4.60	Tea	$1.10
Fried chicken dinner	$11.25	Milk	$1.40
Pork chop dinner	$12.45	Pie	$2.65
Steak dinner	$13.95	Cake	$2.55

1. When Elina went to lunch at Lois's, she ordered soup, a salad, and a cup of tea. How much did her lunch cost?

2. Carl took his afternoon break at Lois's. He had a piece of pie and a cup of coffee. How much was he charged?

3. Mr. Cheung took his assistant Marissa out to lunch at Lois's. Marissa had the pork chop dinner and a cup of coffee. Mr. Cheung had the steak, a cup of coffee, and a piece of cake. What was the total cost for their two lunches?

4. David had a big appetite when he went to Lois's. He ordered soup, a salad, the fried chicken dinner, a piece of pie, and two glasses of milk. How much did his lunch cost?

5. When Kasi and Rajan went to Lois's, they each had the fried chicken dinner. Kasi also had a salad and a cup of tea. Rajan had a cup of coffee and a piece of cake. Together, how much did their meals cost?

Reading Sales Tax Tables

In many states and in some cities, businesses have to charge a tax on anything they sell. This tax is called a **sales tax.** To figure out how much sales tax to charge, businesses are given sales tax tables by the state or local government. The table below is a sample section of a sales tax table that is similar to the ones provided by the government. Although your state sales tax may be higher or lower than the tax shown on this table, every table is set up like the one shown below. In one column you must find the price of the item or items sold. Printed next to the total cost of the sale is the amount of sales tax that must be added. This table shows the sales tax on amounts from 1¢ to $5.91.

Amount	Tax	Amount	Tax	Amount	Tax
$0.01 to 0.08 -	0¢	$0.92 to 1.08 -	6¢	$1.92 to 2.08 -	12¢
0.09 to 0.24 -	1¢	1.09 to 1.24 -	7¢	2.09 to 2.24 -	13¢
0.25 to 0.41 -	2¢	1.25 to 1.41 -	8¢	2.25 to 2.41 -	14¢
0.42 to 0.58 -	3¢	1.42 to 1.58 -	9¢	2.42 to 2.58 -	15¢
0.59 to 0.74 -	4¢	1.59 to 1.74 -	10¢	2.59 to 2.74 -	16¢
0.75 to 0.91 -	5¢	1.75 to 1.91 -	11¢	2.75 to 2.91 -	17¢
$2.92 to 3.08 -	18¢	$3.92 to 4.08 -	24¢	$4.92 to 5.08 -	30¢
3.09 to 3.24 -	19¢	4.09 to 4.24 -	25¢	5.09 to 5.24 -	31¢
3.25 to 3.41 -	20¢	4.25 to 4.41 -	26¢	5.25 to 5.41 -	32¢
3.42 to 3.58 -	21¢	4.42 to 4.58 -	27¢	5.42 to 5.58 -	33¢
3.59 to 3.74 -	22¢	4.59 to 4.74 -	28¢	5.59 to 5.74 -	34¢
3.75 to 3.91 -	23¢	4.75 to 4.91 -	29¢	5.75 to 5.91 -	35¢

Suppose that you are a cashier in a restaurant or a store and you want to find the tax due on a $3.30 purchase. According to the table, the tax on an amount from $3.25 to $3.41 is 20¢; so the tax on $3.30 is 20¢.

Use the table to answer the following questions.

1. How much sales tax is due on each of the following amounts?

a. $1.45 **b.** $0.80

c. $3.60 **d.** $4.10

e. $5.40 **f.** $2.10

g. $3.19 **h.** $4.32

i. $5.06 **j.** $5.89

To find the sales tax for an amount larger than $5.91, first multiply the dollar amount by $0.06. Then add the tax for the remaining amount from the first section of the table on page 134.

EXAMPLE Find the tax on $7.54.

7 × $0.06 = $0.42	STEP 1	Multiply the dollar amount, 7, by $0.06.
+ 0.03		
$0.45	STEP 2	Look up $0.54 on the tax table. The tax on $0.54 is 3¢. Add $0.42 and $0.03.

2. Find the tax on Elina's lunch (problem 1, page 133).

3. What was Elina's total lunch bill including tax?

4. How much change should Elina get back from $15?

5. Find the tax on Carl's pie and coffee (problem 2, page 133).

6. How much tax did Mr. Cheung have to pay on the bill for his and his assistant's lunches (problem 3, page 133)?

7. Mr. Cheung left a $6 tip. How much did he pay altogether for his and his assistant's lunches including tax and tip?

8. Find the tax on David's lunch (problem 4, page 133).

9. How much change should David receive from $30?

10. Find the tax on the lunch bill for Kasi and Rajan (problem 5, page 133).

11. Rajan paid for his and Kasi's meals. How much change did he get back from $40?

Checking Your Change with Sales Receipts

On this page and the next are copies of sales receipts. On each receipt, a number followed by + indicates the price of an item. A number followed by TX means tax. A number followed by TL means a total.

Use the sales receipts to answer the following questions. People who like to get rid of loose change should pay particular attention to the second problem for each receipt.

Frank's Foods
2930 Broadway
$06.99 +
$01.56 +
$01.27 +
$02.82 +

$12.64 TL
$01.01 TX

$13.65 TL

1. How much change should you receive from $20?

2. How much change would you receive if you give the clerk a $10 bill, a $5 bill, and 15¢ in change?

Doug's Drugs
4 East 110th
$02.69 +
$03.56 +
$00.98 +

$07.23 TL
$00.58 TX

$07.81 TL

3. How much change should you receive from $10?

4. If you paid with a $5 bill, three $1 bills, and a penny, how much change would you have gotten back?

Bonnie's Books
854 West George
$07.95 +
$14.99 +

$22.94 TL
$1.84 TX

$24.78 TL

5. How much change should you receive from $30?

6. If you paid with a $20 bill, a $5 bill, and 3 pennies, how much change would you have received?

Hal's Hardware
5 East Third St.

$09.59 +
$16.44 +
$18.27 +

$44.30 TL
$03.54 TX

$57.74 TL

7. How much change should you get back from $100?

8. If you paid with three $20 bills and three quarters, how much change would you have received?

Paula's Pastry
844 W. Oakdale

$03.05 +
$06.15 +
$02.95 +

$12.15 TL
$00.97 TX

$13.12 TL

9. How much change should you get back from $20?

10. If you paid with a $10 bill, a $5 bill, two nickels, and two pennies, how much change would you have gotten back?

Colette's Clothes
333 East 49th St.

$27.29 +
$04.85 +
$33.99 +
$16.50 +

$82.63 TL
$06.61 TX

$89.24 TL

11. How much change should you get back from $100?

12. If you paid with four $20 bills, a $10 bill, two dimes, and a nickel, how much change would you have gotten back?

Using a Calorie Chart

Everything we eat has a certain number of calories. People who want to lose weight often have to cut down on the number of calories they take into their bodies each day. A doctor can tell you how many calories you need to stay healthy and lose weight at the same time. Dieters generally use a calorie chart to keep track of how many calories they are taking in.

Below is a short calorie chart. Use this chart to answer the following questions.

Item	Quantity	Calories
chili con carne	1 small bowl	250
jelly doughnut	1	250
frankfurter	1	125
frankfurter roll	1	125
hamburger, fried	2-ounce patty	225
hamburger roll	1	250
milk shake	1	350
pretzels	5 small sticks	20
skim milk	8-ounce glass	90
cola	6 ounces	75
ginger ale	6 ounces	75
coffee with cream and sugar	1 cup	80

1. Martha stopped at the Doughnut Barn for a snack. She had two jelly doughnuts and coffee with cream and sugar. How many calories did she take in?

2. For her breakfast Marissa took in 600 calories, and for lunch, 715 calories. If she sticks to a limit of 2,400 calories per day, how many calories is she allowed for dinner?

3. Jeff stopped at the Come On Inn for breakfast. He had three jelly doughnuts, two 8-ounce glasses of skim milk, and one coffee with cream and sugar. How many calories did Jeff take in?

4. Before going home, Jeff went to Herb's Hamburgers where he ate two hamburgers on hamburger rolls and drank a milk shake. How many calories did Jeff take in at Herb's?

5. For lunch Alicia had a small bowl of chili and a cola. How many calories did she take in for lunch?

6. Mei's breakfast contained 280 calories. If she had a snack of 235 calories before dinner, and she is allowing herself only 2,000 calories a day, how many calories can Mei take in at dinner?

7. At a baseball game, John had two frankfurters on rolls, a box of 50 pretzels, and four colas. How many calories did John take in at the game?

8. Helen had a breakfast of 475 calories. For lunch she had two hamburgers without rolls and a glass of ginger ale. If she limits herself to 2,000 calories a day, how many calories can there be in her dinner?

9. During his morning break, Felix had two jelly doughnuts and a cup of coffee with cream and sugar. During his afternoon break, he had a milk shake. How many calories did Felix take in during his breaks?

10. From noon to 6:00 P.M., Sergio had a bowl of chili, two colas, a frankfurter on a roll, and a cup of coffee with cream and sugar. How many calories did he take in during this time?

Reading Paycheck Stubs

Below is a copy of a weekly paycheck stub from Addie Strand's paycheck. It shows her gross pay (before deductions) and her net pay (after deductions). It also shows the amounts withheld for FICA (social security) and various taxes for both the pay period shown and the year to date.

Use this check stub to answer the following questions.

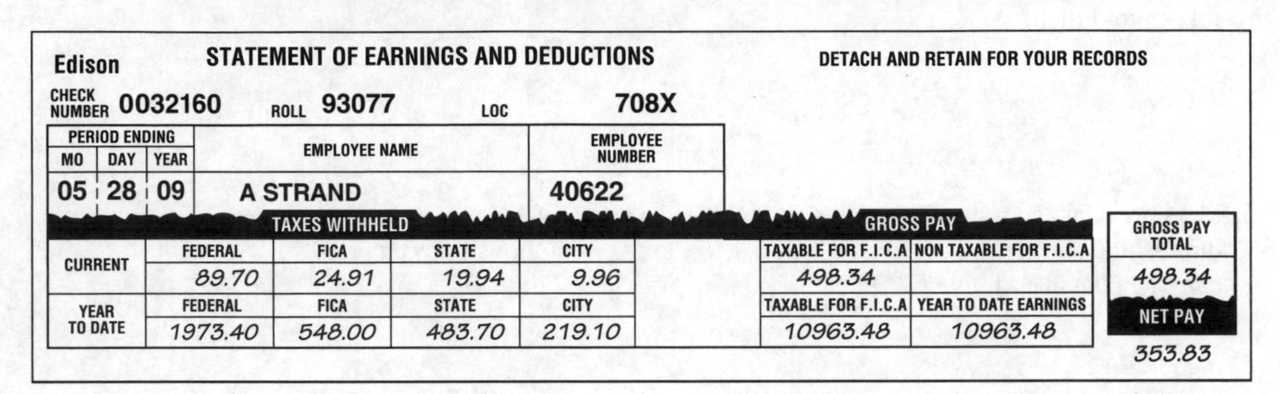

Edison — STATEMENT OF EARNINGS AND DEDUCTIONS — DETACH AND RETAIN FOR YOUR RECORDS

CHECK NUMBER 0032160 ROLL 93077 LOC 708X

PERIOD ENDING MO	DAY	YEAR	EMPLOYEE NAME	EMPLOYEE NUMBER
05	28	09	A STRAND	40622

TAXES WITHHELD

	FEDERAL	FICA	STATE	CITY
CURRENT	89.70	24.91	19.94	9.96
YEAR TO DATE	1973.40	548.00	483.70	219.10

GROSS PAY

TAXABLE FOR F.I.C.A	NON TAXABLE FOR F.I.C.A
498.34	
TAXABLE FOR F.I.C.A	**YEAR TO DATE EARNINGS**
10963.48	10963.48

GROSS PAY TOTAL
498.34
NET PAY
353.83

1. What is Ms. Strand's gross pay each week?

2. What is Ms. Strand's net pay each week?

3. How much has been withheld to date for federal tax?

4. How much has been withheld to date for social security?

5. How much has been withheld to date for state tax?

6. What is Ms. Strand's gross pay to date?

7. If Ms. Strand makes the same amount for 52 weeks in the year, what is her yearly gross pay?

8. If she has the same deductions all year, what is Ms. Strand's net pay for the year?

9. There are 30 more pay periods for Ms. Strand for the year indicated on the paycheck stub. How much take-home pay will Ms. Strand receive for the rest of the year?

10. The same amount is withheld each week from her check. Find the total amount of federal tax that is withheld from Ms. Strand's pay in one year.

11. Ms. Strand budgets $165 a week for food for her children and herself. How much is left from her take-home pay each week after paying for food?

12. Ms. Strand's house mortgage amounts to $107.50 per week. How much does she have left from her weekly net pay after paying for food and the mortgage?

13. Ms. Strand's car payments are $38.60 per week. How much does she have left from her weekly net pay after paying for food, the mortgage, and the car?

14. Which of the following approximates the amount Ms. Strand has left each *month* for clothes, utilities, savings, entertainment, and miscellaneous expenses?

a. $345
b. $295
c. $215
d. $175

Using a Table to Look Up Check Cashing Rates

If you cash a check at a private check cashing establishment rather than at a bank or at the company where you work, you will be charged a fee. This fee is deducted from the amount of the check.

The table below shows the fees to be charged at a certain check cashing establishment for checks from $1.00 to $204.99.

Honest John's Check Cashing

$1.00 to 14.99 - 20¢	$75.00 to 79.99 - $0.85	$140.00 to 144.99 - $1.50
15.00 to 19.99 - 25¢	80.00 to 84.99 - 0.90	145.00 to 149.99 - 1.55
20.00 to 24.99 - 30¢	85.00 to 89.99 - 0.95	150.00 to 154.99 - 1.60
25.00 to 29.99 - 35¢	90.00 to 94.99 - 1.00	155.00 to 159.99 - 1.65
30.00 to 34.99 - 40¢	95.00 to 99.99 - 1.05	160.00 to 164.99 - 1.70
35.00 to 39.99 - 45¢	100.00 to 104.99 - 1.10	165.00 to 169.99 - 1.75
40.00 to 44.99 - 50¢	105.00 to 109.99 - 1.15	170.00 to 174.99 - 1.80
45.00 to 49.99 - 55¢	110.00 to 114.99 - 1.20	175.00 to 179.99 - 1.85
50.00 to 54.99 - 60¢	115.00 to 119.99 - 1.25	180.00 to 184.99 - 1.90
55.00 to 59.99 - 65¢	120.00 to 124.99 - 1.30	185.00 to 189.99 - 1.95
60.00 to 64.99 - 70¢	125.00 to 129.99 - 1.35	190.00 to 194.99 - 2.00
65.00 to 69.99 - 75¢	130.00 to 134.99 - 1.40	195.00 to 199.99 - 2.05
70.00 to 74.99 - 80¢	135.00 to 139.99 - 1.45	200.00 to 204.99 - 2.10

Find the fee for cashing checks for each of the following amounts.

1. $43.89

2. $80.10

3. $98.50

4. $153.99

5. $104.85

6. $203.57

7. $76.99

8. $119.60

9. $134.95

10. $187.92

11. Hector took his paycheck of $137.50 to Honest John's. How much cash should Hector receive?

12. Sonia cashed her check for $193.65 at Honest John's. She also paid her gas and electricity bill for $42.47. How much did she take home from Honest John's?

13. Doreen cashed her check for $142.75 and paid her phone bill for $36.58 at Honest John's. How much cash did she take home?

14. Tony cashed his check for $198.70 and his wife's check for $168.45 at Honest John's. If he had to pay a separate charge for each check, how much did Tony take home for himself and his wife?

15. Adnan cashed his $184.00 tax refund and paid his $34.65 phone bill and $29.42 gas and electricity bill at Honest John's. How much did he take home?

16. Frank cashed his check from his part-time job for $174.29 at Honest John's. Frank's brother Pete went with him to collect the $50.00 Frank had borrowed a month before. How much did Frank take home after paying the money he owed Pete?

17. Andrea cashed her paycheck for $182.60 and her rebate of $25.00 for her new TV at Honest John's. She had to pay a separate fee for each check. How much money did she take home?

Understanding Tax Statements

At the beginning of every year, employers give their employees a wage and tax statement (W-2 Form), which must be attached to everyone's federal and state income tax forms. The wage and tax statement gives the following information:

1. the total amount of the employee's salary before deductions were taken out, called **gross salary** (shown in box 2)
2. the amount deducted for federal income tax (shown in box 1)
3. the amount deducted for FICA, or social security (shown in box 3)
4. the amount deducted for state income tax (shown in box 6)
5. the amount deducted for city income tax (shown in box 9)

The year shown in the upper right-hand corner of the statement is the year in which the salary and taxes were paid.

The wage and tax statement below shows how much Adam Smith earned in 2009 while he was working at Positive Systems in New York.

Use this statement to answer the following questions.

POSITIVE SYSTEMS
150 NASSAU ST.
NEW YORK, N.Y.

Wage and Tax Statement **2009**

Copy C For employee's records

Type or print EMPLOYER'S Federal identifying number, name, address and ZIP code above.

FEDERAL INCOME TAX INFORMATION		SOCIAL SECURITY INFORMATION		STATE OR LOCAL TAX INFORMATION		
1 Federal income tax withheld	2 Wages, tips and other compensation	3 FICA employee tax withheld	4 Total FICA wages	6 State Tax withheld	7 Wages paid	8 Name of State
5,271.32	29,285.00	1,698.53	29,285.00	1,171.42		NEW YORK
Type or print EMPLOYEE'S social security no., name, and address including ZIP code below		5 Uncollected employee FICA tax on tips		9 City Tax withheld	10 Wages paid	11 Name of City
				409.99		NEW YORK

123–45–6789

SMITH, ADAM
950 W. 57th ST.
N.Y., N.Y. 10019

Employee No.	Form No. State	Form No. City or Other	Was employee covered by a qualified pension plan etc.?
	IT–2102	NYC–2	Yes ☒ No ☐

OTHER INFORMATION (SEE CIRCULAR E)

Cost of group term life insurance included in box 2.7.10	Excludable sick pay included in box 2.7.10		Contribution to individual employee retirement account

From **W-2** AN "X" in the upper left corner indicates this is a corrected form. 22 1766474 This information is being furnished to the Internal Revenue Service and appropriate State officials.

1. How much did Mr. Smith make during 2009 before deductions?

2. How much was deducted from Mr. Smith's wages for FICA (social security)?

3. How much was withheld from Mr. Smith's wages for federal income tax?

4. How much was withheld from Mr. Smith's wages for state income tax?

5. How much was withheld from Mr. Smith's wages for city income tax?

6. What were the total deductions taken from Mr. Smith's wages in 1999?

7. What was Mr. Smith's net income (take-home pay) for 1999?

8. To the nearest ten dollars, what was Mr. Smith's average monthly take-home pay?

9. Mr. Smith and his wife pay $572.60 a month for rent. How much do they spend on rent in one year?

10. How much is left from Mr. Smith's yearly take-home income after rent is paid?

11. The average food bill for Mr. and Mrs. Smith is $118 per week. How much do they spend on food in one year? (1 year = 52 weeks)

12. How much is left from Mr. Smith's take-home pay when both rent and food are deducted?

13. The average monthly gas and electricity bills for the Smiths is $41.50. How much do they pay for gas and electricity in one year?

14. How much is left from Mr. Smith's take-home pay after rent, food, gas, and electricity are deducted?

Figuring the Cost of Electricity

The chart below shows the average costs for operating various electrical appliances. Charts like this one are distributed by power companies to their customers, although rates are different in different cities. Using a chart like this helps people understand and keep down their electricity bills.

Use this chart to answer the following questions. (Use 1 month = 30 days.)

Appliance	Average Cost	Appliance	Average Cost
Refrigerator (single door)		**Blender**	45 uses for 1¢
12 cu ft, manual defrost	21¢ a day	**Mixer (hand)**	15 uses for 1¢
Refrigerator/Freezer (two door)		**Television**	3¢ per hour
14 cu ft, cycle defrost	33¢ a day	**Lightbulb**	
14 cu ft, frostless	39¢ a day	100-watt	1 hour for 1¢
Freezer		60-watt	2 hours for 1¢
15 cu ft chest, manual defrost	36¢ a day	**Room air conditioner**	16¢ per hour
16 cu ft upright, frostless	63¢ a day	**Electric blanket**	9¢ per night
Dishwasher	7¢ per use	**Clock**	15¢ per month
Toaster	4 slices for 1¢	**Vacuum cleaner**	5¢ per hour
Hair dryer	13¢ per hour	**Washing machine**	2¢ per load
		Clothes dryer	28¢ per load
		Iron	5¢ per hour
		Computer	3¢ per hour

1. How much does it cost to run a television for 3 hours?

2. If Lynne watches her television an average of 3 hours every night, how much does it cost her to watch television in 1 month?

3. How much does it cost to keep a 12-cubic-foot, single-door refrigerator running for 1 month?

4. How much does it cost to use an iron for 8 hours?

5. Find the cost of burning four 60-watt bulbs for 6 hours.

6. How much would it cost to burn four 60-watt light bulbs for 6 hours every day for a month?

7. To cool his electronics repair shop during the summer months, Mr. Rosa uses his room air conditioner 9 hours a day. How much does it cost him to run his air conditioner for 9 hours?

8. If Mr. Rosa uses his air conditioner 9 hours every day for a month, how much will it cost him?

9. How much does it cost to toast 100 slices of bread?

10. The Frey family includes three teenage girls and one 8-year-old boy. Following is an account of how much electricity the Frey family used in one month. Figure out the cost for each appliance and the monthly electricity bill.

Appliance	Time Use	Monthly Cost
14 cu ft frostless refrig/freezer	all the time	______
television	2 hours per day	______
ten 60-watt lightbulbs	6 hours per day	______
electric clock	all the time	______
iron	4 hours a month	______
vacuum cleaner	10 hours a month	______
computer	4 hours per day	______
hair dryer	1 hour per day	______
	Total	______

Reading Time Schedules

NEW YORK/BOSTON

	Leave	Arrive	Flight No.	Stops or Via	Meals	Equip.	Freq.
To Boston	7:00a L	7:53a	308	NON-STOP	☕	727	ExSaSu
F $319.00	7:15a K	8:09a	10	NON-STOP	☕	D10	Daily
Y $143.00	8:35a L	9:32a	496	NON-STOP	☕	727	ExSu
	9:45a L	10:37a	360	NON-STOP		727	Daily
	11:50a L	12:42p	606	NON-STOP	🍸	727	Daily
	1:20p L	2:10p	82	NON-STOP	🍸	727	Daily
	4:05p L	4:59p	579	NON-STOP	🍸	727	Daily
	5:10p K	6:07p	186	NON-STOP	🍸	707	Daily
	5:40p K	6:43p	126	NON-STOP	🍸	D10	Daily
	7:30p L	8:24p	228	NON-STOP	🍸	727	Daily
	8:30p L	9:19p	276	NON-STOP	🍸	727	ExSa
	9:30p L	10:19p	598	NON-STOP	🍸	727	ExSa
	10:00p K	10:57p	118	NON-STOP	🍸	707	Daily
	1:10a K	2:01a	4	NON-STOP	🍸	707	ExSu
From Boston	7:00a	7:52a L	577	NON-STOP	☕	727	ExSaSu
	7:35a	8:29a K	117	NON-STOP	☕	707	Daily
	8:40a	9:32a L	599	NON-STOP	☕	727	ExSu
	10:35a	11:27a L	85	NON-STOP	🍸	727	Daily
	11:25a	12:17p L	369	NON-STOP	🍸	727	Daily
	1:20p	2:11p L	249	NON-STOP	🍸	727	Daily
	2:45p	3:41p K	187	NON-STOP	🍸	707	Daily
	4:05p	4:57p L	261	NON-STOP	🍸	727	ExSa
	5:15p	6:07p L	511	NON-STOP	🍸	727	Daily
	6:50p	7:43p L	341	NON-STOP	🍸	727	ExSa
	6:50p	7:43p L	329	NON-STOP	🍸	727	Sa
	8:25p	9:15p L	545	NON-STOP	🍸	727	ExSa
	9:00p	9:58p K	95	NON-STOP	🍸	707	Daily

The schedule above is for one airline's flights between the New York City airports and Boston. Following is a list of the meanings of the abbreviations used in the schedule.

F—first class ticket (one way)
Y—coach ticket (one way)
a—morning (A.M.)
p—afternoon/evening (P.M.)
K—Kennedy Airport
L—LaGuardia Airport

Flight No.—flight number
Equip.—type of airplane
Freq.—how often the flight leaves (what days)
Ex—except

Use the New York/Boston schedule to answer the following questions.

1. At what time does flight number 606 leave from New York?

2. At what time does flight number 606 arrive in Boston?

3. How long does flight number 606 take to get to Boston? (Remember to borrow 60 minutes.)

4. How long does flight 341 take to get from Boston to New York?

5. Does flight 228 go to Boston every day?

6. Does flight 577 go to New York every day?

7. How long does the earliest flight leaving from Kennedy Airport to Boston take to get to Boston?

8. Mr. Parker paid $12.00 for a bus ride from midtown New York to LaGuardia Airport and $32.50 for a taxi ride from Boston's airport to downtown Boston. If he buys a coach ticket, how much does the trip from midtown New York to downtown Boston cost Mr. Parker?

9. If Mr. Parker pays the same fares for his return, what is the round-trip cost from midtown New York to downtown Boston and back to midtown New York?

10. Ms. Ramos took flight number 360 to Boston and returned the same day on flight number 511. How much time did she have in Boston? (Find how much time before noon and how much after noon.)

11. How much is round-trip coach fare between New York and Boston?

12. How much is round-trip first-class fare between New York and Boston?

13. How much does a passenger save on round-trip fare between New York and Boston by going coach rather than first class?

Using the Cost Formula

A **formula** is a mathematical instruction in which letters represent numbers. The cost formula is $c = nr$ in which c represents total cost, n represents the number of units, and r represents the rate, or cost, for each unit. When two letters are written next to each other in a formula, the numbers those letters represent should be multiplied together. In words, the cost formula means "cost equals number of units multiplied by rate per unit."

EXAMPLE **Sarah bought three boxes of pasta that each cost $1.45. Find the total cost.**

$c = nr$
$c = 3 \times \$1.45$
$c = \$4.35$

STEP 1 Substitute 3 for n and $1.45 for r in the cost formula.

STEP 2 Multiply $1.45 by 3. The total cost is $4.35.

Use the cost formula to solve each problem.

1. Max is a manager at a hardware store. He orders 60 screwdriver sets from his supplier. The supplier charges $24 for each set. How much will Max have to pay the supplier for the order?

2. If Max charges his customers $28.90 for each set of screwdrivers, how much profit can he make on the order if he sells all the sets?

3. From the same supplier, Max orders 50 hammer sets that cost $17.50 each. What is the total cost of Max's order for screwdrivers and hammers?

4. Lei works at the summer soccer program for the parks department in her town. She decides to buy 20 soccer balls at a cost of $17 each and 10 better quality soccer balls at a cost of $32 each. What is the total cost of the soccer balls?

5. Lei also orders 120 soccer jerseys that cost $29 each and 120 pairs of soccer shorts that cost $33 each. Find the total cost of the jerseys and shorts. (**Hint:** First find the combined cost of a jersey and a pair of shorts.)

6. Lei's purchasing budget for the summer is $10,000. How much is left after she buys the soccer balls, the jerseys, and the shorts?

7. At Good Buys, DVDs of 1980s movies that usually cost $15 each are now on sale for $8 each. Glenn wants to purchase two dozen DVDs for the local library. How much can he save if he buys the DVDs on sale?

The cost formula ($c = nr$) shows the relationship among three numbers—cost, number, and rate. The cost formula can be written three ways. If you know the cost and the number of items, use $r = c/n$ to find the rate. In words, this formula means, "unit rate equals total cost divided by number of items."

If you know the total cost and the unit rate, use $n = c/r$ to find the number of items. In words, this formula means, "number of items equals total cost divided by unit rate."

Use the correct form of the cost formula to solve each problem.

8. Laura is in charge of buying furniture for a community center. The total cost for 120 new folding chairs is $2,280. What is the price per chair?

9. Alex installed new carpet in an office. The area of the office floor is 180 square yards. The total cost of the carpet including installation was $4,860. Find the cost per square yard of the carpet.

10. Amor bought baseball caps with her company's logo printed on each cap to give to everyone at the annual company picnic. The caps cost $13 each, and Amor paid a total of $780 for the caps. How many caps did she buy?

11. Amor is thinking of buying T-shirts printed with the company logo. If the T-shirts cost $16 each, how many can she buy for $1,000?

12. Carla works in a pet supply store. She orders 180 thirty-pound bags of dry dog food. If the order totals $7,020, what is the price for each bag?

13. Carla also orders 16-pound bags of dry cat food for $19 each. The supplier charges Carla $4,750 for the order of cat food. How many bags did she receive?

14. Along with the bags of dog food and the bags of cat food, Carla orders 240 cans of cat food that cost $1.25 each. Find the total cost of her order.

Using the Distance Formula

The distance formula is $d = rt$ in which d represents distance, r represents the rate (usually measured in miles per hour), and t represents the time. In words, the distance formula means "distance equals rate multiplied by time."

EXAMPLE **Larry drove for three hours at a rate of 65 mph. Find the total distance that he drove.**

$d = rt$
$d = 65 \times 3$
$d = 195$ miles

STEP 1 Substitute 65 for r and 3 for t in the distance formula.

STEP 2 Multiply 65 by 3. Larry drove 195 miles.

Use the distance formula to solve each problem.

1. An airplane maintained an average speed of 436 miles per hour. How far did the plane travel during a four-hour flight?

2. Melissa and her neighbor try to take a long walk every Saturday morning. For the first hour they walk at an average speed of 3 mph through a hilly part of town. Then they walk for two hours at an average speed of 4 mph in a less hilly part of town. How many miles do they walk each Saturday?

3. Juan drove his delivery van for one hour in heavy traffic at an average speed of 12 mph. Then he drove for four hours at an average speed of 64 mph. Finally he drove for two hours on country roads at an average speed of 42 miles per hour. What total distance did Juan drive that day?

4. Joseph and his wife Janet both drive long distances to work. To get to his job, Joseph drives two hours at an average speed of 38 miles per hour. Janet also drives two hours at an average speed of 43 miles an hour to get to her job. Janet's job is how much farther from home than Joseph's?

5. A train traveled through the night at an average speed of 67 miles per hour. How far did the train travel in six hours?

6. In four seconds how far can a golf ball travel if its speed is 237 feet per second?

7. A high-speed train in France can travel at a maximum of 357 miles per hour. A high-speed train in the United States can travel at a maximum of 150 miles per hour. In two hours, how much farther can a French high-speed train travel than a U.S. high-speed train?

The distance formula ($d = rt$) shows the relationship among three numbers—distance, rate, and time. The distance formula can be written three ways. If you know the distance and the time, use $r = d/t$ to find the rate. In words, this formula means, "rate equals total distance divided by the time."

If you know the distance and the rate, use $t = d/r$ to find the time. In words, this formula means, "time equals the distance divided by the rate."

Use the correct form of the distance formula to solve each problem.

8. Jack makes extra money driving cars to distant locations. He has to deliver a car to Florida for a couple who are retiring there. The town in Florida is 675 miles from where Jack will pick up the car. If he drives at an average speed of 45 miles an hour, how many hours will it take him to get to his destination?

9. Samly teaches drivers' education four nights a week. Each evening he spends two hours with a student doing practice driving. One week the odometer (mileage gauge) on the teaching car showed that the car had been driven 232 miles during practice driving. Find the average speed that the car was driven that week.

10. Peter and his wife Carolina both rode in a charity bicycle ride that was 60 miles long. Peter's average speed was 20 mph. Carolina is not an experienced rider, and her average speed was only 12 mph. How much more time did Carolina need to complete the ride than Peter needed?

11. A fighter jet was able to fly 4,260 miles in three hours. What was the average speed of the jet?

12. Jana has to drive 768 miles to get back to school. What average speed will she have to drive to get there in 12 hours?

13. In fact, Jana needed 15 hours to get back to school. To the nearest *ten,* what average speed did she drive?

Renting a Car

Many car rental agencies charge different rates for different types of cars. Your car rental bill will also depend on which days of the week you rent the car as well as the number of days you keep it. Notice that Ralph's Rent-a-Car (in the ad shown below) offers unlimited free mileage. That means that there is no charge for the distance that you drive the car.

Use the ad below to answer the following questions.

Ralph's Rent-a-Car Unlimited Free Mileage			
Model	**Monday to Thursday (Per Day)***	**Fri., Sat., Sun. & Holidays (Per Day)***	**Weekly, Any 7 Consecutive Days**
Compact	$23.95	$36.95	$165.95
Deluxe Compact	$28.95	$39.95	$187.95
Intermediate	$32.95	$42.95	$199.95
Standard	$36.95	$45.95	$211.95

*You will be charged the full day rate regardless of the time of day you pick up and return the car.

1. On Friday morning, Morris rented a deluxe compact from Ralph's Rent-a-Car. He returned it Sunday night. How much was he charged?

2. Mr. and Mrs. Colon rented a standard car from Ralph's. They picked up the car on Wednesday and returned it the next Tuesday. How much did they have to pay at Ralph's?

3. How much would it cost to rent an intermediate size car from Sunday night until noon on Thursday?

4. How much does it cost to rent a compact car for the 3-day Labor Day weekend?

5. Mr. Slater rented a deluxe compact car Friday night and returned it on Sunday. During that time, he filled the tank three times for the following prices: $20.50, $16.90, and $19.15. How much did he spend on gasoline and car rental for that weekend?

6. Mr. and Mrs. Sosa rented a standard size car from Ralph's on Sunday evening and returned it during the day on Thursday. If they bought 40 gallons of gas at $2.85 a gallon, how much did they pay for the car including gas and rental fee?

Posttest B

This posttest has a multiple-choice format much like the GED and other standardized tests. Take your time and work each problem carefully. Circle the correct answer to each problem. When you finish, check your answers at the back of the book.

1. Subtract 6,894 from 70,293.

- **a.** 63,399
- **b.** 64,109
- **c.** 64,799
- **d.** 64,979

2. What is the sum of 1,296 and 3,978?

- **a.** 6,154
- **b.** 5,274
- **c.** 5,184
- **d.** 4,934

3. The CN Tower in Toronto is 1,821 feet high. What is the value of the digit 8 in the number 1,821?

- **a.** 8
- **b.** 80
- **c.** 800
- **d.** 8000

4. Silvia's new sofa cost $783 including tax and delivery charges. She made a down payment of $135. Find the balance due on the sofa.

- **a.** $588
- **b.** $598
- **c.** $618
- **d.** $648

5. Silvia, in the last problem, paid off the balance in 12 equal monthly payments. How much did she pay each month?

- **a.** $44
- **b.** $48
- **c.** $52
- **d.** $54

6. In the number 82,564, which digit is in the thousands place?

- **a.** 8
- **b.** 2
- **c.** 5
- **d.** 6

7. $4{,}856 \div 8 =$

- **a.** 670
- **b.** 667
- **c.** 607
- **d.** 67

8. $204{,}070 - 8{,}329 =$

- **a.** 196,951
- **b.** 195,741
- **c.** 204,359
- **d.** 212,399

9. $487 \times 9 =$

- **a.** 3,893
- **b.** 3,913
- **c.** 4,383
- **d.** 4,423

10. $59 + 3 + 47 + 1 =$

- **a.** 100
- **b.** 110
- **c.** 120
- **d.** 130

11. $10{,}000 - 9{,}326 =$

a. 1,034
b. 994
c. 784
d. 674

12. Mr. and Mrs. Alvarez pay $677 a month for rent. How much rent do they pay in one year?

a. $6,770
b. $7,190
c. $8,124
d. $9,034

13. $544 / 32 =$

a. 16
b. 17
c. 21
d. 22

14. $\frac{8{,}235}{27} =$

a. 350
b. 335
c. 325
d. 305

15. $56 + 2{,}043 + 837 =$

a. 2,936
b. 4,126
c. 8,440
d. 16,013

16. $(74)(358) =$

a. 26,492
b. 25,942
c. 24,392
d. 21,082

17. $56{,}032 - 9{,}547 =$

a. 65,579
b. 55,479
c. 46,485
d. 45,395

18. $16 \cdot 4{,}039 =$

a. 60,544
b. 61,414
c. 62,864
d. 64,624

19. The balance due on Mr. and Mrs. Chan's mortgage is $59,927. Round the balance to the nearest *thousand* dollars.

a. $57,000
b. $58,000
c. $59,000
d. $60,000

20. Herbert works four days a week as a traveling salesman. On Monday he drove 417 miles, on Tuesday he drove 308 miles, on Wednesday he drove 289 miles, and on Thursday he drove 318 miles. Altogether, how many miles did he drive those days?

a. 725
b. 1,014
c. 1,043
d. 1,332

21. In the last problem, find the average number of miles that Herbert drove each day.

a. 293
b. 333
c. 343
d. 373

22. In a week Herbert often drives 1,500 miles. The total distance that he drove on the four days listed in problem 20 is how many miles less than 1,500?

a. 32
b. 68
c. 132
d. 168

23. $3{,}741 + 29{,}634 + 8{,}025 =$

a. 41,400
b. 44,050
c. 75,069
d. 147,294

24. $28{,}700 \div 70 =$

a. 4,101
b. 4,100
c. 410
d. 41

25. Round each number in the problem 8,672 + 17,386 to the nearest *thousand.* Then add.

a. 24,000
b. 25,000
c. 26,000
d. 27,000

26. In a recent year, the United States imported 245,296 passenger cars from Germany. Round the number of imported cars to the nearest *ten thousand.*

a. 200,000
b. 240,000
c. 250,000
d. 260,000

27. $(293)(1{,}000) =$

a. 2,930
b. 29,300
c. 293,000
d. 2,930,000

28. In a town election, three candidates ran for city council. The winner received 1,284 votes. The second-place candidate received 996 votes, and the third-place candidate received 381 votes. What was the total number of votes cast?

a. 1,861
b. 2,371
c. 2,551
d. 2,661

29. In the problem 12,827 – 8,549, round each number to the nearest *thousand.* Then subtract.

a. 3,000
b. 4,000
c. 5,000
d. 6,000

30. From her check, Carmen's employer deducts $167 for federal tax, $44.68 for social security, and $39.44 for state tax. Find the total of these deductions.

a. $290.92
b. $251.12
c. $231.28
d. $211.68

31. Carmen, in the last problem, gets paid every two weeks. She earns a *weekly* gross salary of $480.00. What is her *weekly* take-home pay?

a. $354.44
b. $325.56
c. $296.44
d. $286.56

32. What is the quotient of 49,634 divided by 13 to the nearest *hundred?*

a. 800
b. 3,600
c. 3,700
d. 3,800

33. Round both 904 and 7,308 to the nearest *thousand.* Then multiply.

a. 70,000,000
b. 63,000,000
c. 7,000,000
d. 6,300,000

34. Which of the following is the *least* amount that is sufficient to pay for 5 gallons of paint that cost $18.79 per gallon?

a. $75
b. $85
c. $90
d. $100

35. One month a travel agency sold 194 tickets for a total of $54,958. Which of the following is closest to the average price of one ticket?

a. $180
b. $280
c. $380
d. $440

POSTTEST B CHART

If you missed more than one problem on any group below, review the practice pages for those problems. If you had a passing score, redo any problem you missed.

Problem Numbers	Skill Area	Practice Pages
3, 6, 19, 26	place value	6–12
2, 10, 15, 20, 23, 25, 28, 30	addition	15–28
1, 4, 8, 11, 17, 22, 29, 31	subtraction	33–50
9, 12, 16, 18, 27, 33, 34	multiplication	56–78
5, 7, 13, 14, 21, 24, 32, 35	division	84–110

ANSWER KEY

Pages 1–4, Pretest

1. 3	**13.** 115	**25.** 71,225
2. 51,000 sq mi	**14.** 2,029	**26.** 720,000
3. 7	**15.** 868	**27.** 390 miles
4. 3,500	**16.** 13,877	**28.** 22,600 feet
5. 98	**17.** 773,647	**29.** 67
6. 4,443	**18.** 2,300	**30.** 806 r 3
7. 80	**19.** $18,900	**31.** 56
8. 31,645	**20.** 439,900	**32.** 730
9. 345,560	**21.** 1,134	**33.** 51
10. 49,900	**22.** 3,222	**34.** 760
11. 1,230 miles	**23.** 4,104	**35.** 200 bundles
12. $36,987	**24.** 25,800	**36.** 23 points

Pages 9–10

1. hundreds and units	**13.** 204,972
2. hundreds and tens	**14.** 339
3. hundreds, tens, and units	**15.** 5,109
4. tens and units	**16.** 13,000
5. 5,600	**17.** 802
6. 7,002	**18.** 34,086
7. 44,900	**19.** 680,000
8. 62,403	**20.** 240,700
9. 820,700	**21.** 1,340,600
10. 2,300,000	**22.** 3,425,100
11. 49,736	**23.** 9,873,000
12. 4,082	**24.** 333,300

Pages 6–8

1. 7	**16.** 0
2. 8	**17.** 700
3. 7,000	**18.** 6
4. 800	**19.** 0
5. 50	**20.** 20
6. 6	**21.** 1
7. 2	**22.** 9
8. 4	**23.** 400
9. 9,000	**24.** 8
10. 200	**25.** 9
11. 0	**26.** 200
12. 4	**27.** 3
13. 5,000	**28.** 9
14. 2	**29.** 8
15. 8	**30.** **d.** 2,536

Pages 11–12

1. 80	**14.** 500	**27.** 63,000
2. 130	**15.** 18,800	**28.** 10,000
3. 260	**16.** 900	**29.** 30,000
4. 5,070	**17.** 6,900	**30.** 140,000
5. 290	**18.** 400	**31.** 80,000
6. 90	**19.** 4,000	**32.** 250,000
7. 14,360	**20.** 8,000	**33.** 170,000
8. 330	**21.** 13,000	**34.** 530,000
9. 980	**22.** 8,000	**35.** 80,000
10. 300	**23.** 236,000	**36.** 1,240,000
11. 3,500	**24.** 9,000	**37.** **c.** $23,000
12. 700	**25.** 27,000	**38.** **d.** $140,000
13. 2,300	**26.** 1,000	**39.** **b.** 210,000

Pages 13–14, Addition Skills Inventory

1. 6,989
2. 86,979
3. 594,988
4. 146
5. 255
6. 241
7. 1,440
8. 864
9. 1,375
10. 702
11. 289
12. 83,789
13. 24,603
14. 1,337,836
15. 185
16. 7,000 + 1,300 + 4,600 = 12,900
17. 30,000 + 32,000 + 49,000 = 111,000
18. 59,000 + 29,000 + 61,000 + 50,000 = 199,000
19. 5,400 + 6,100 + 7,800 + 2,400 = 21,700
20. 82 games
21. 174,857 votes
22. 6,400 + 6,000 + 6,800 + 7,000 = 26,200
23. $54.49

Pages 15–16

1. 9 8 11 5 8 6 5 12 13 9
2. 9 11 10 10 16 2 6 9 14 5
3. 11 4 7 13 9 12 7 3 15 7
4. 6 6 14 9 9 5 10 8 1 8
5. 12 7 8 16 11 11 6 10 13 14
6. 4 17 8 10 7 5 8 4 4 10
7. 8 12 10 12 3 13 7 14 3 2
8. 11 5 12 9 15 6 6 7 14 2
9. 9 13 13 1 10 9 10 18 7 11
10. 12 3 17 11 15 8 15 4 0 16
11. 6 2 12 5 11 5 13 6 9 8
12. 11 10 7 14 8 8 10 12 13 17
13. 5 6 7 10 6 7 6 10 8 15
14. 9 9 11 12 4 8 11 9 12 5
15. 10 13 6 9 16 10 12 9 1 11
16. 9 5 7 8 14 10 2 2 11 16
17. 12 9 3 10 4 14 7 8 18 4
18. 8 11 6 7 17 12 5 13 3 4
19. 8 16 9 9 11 7 3 15 10 15
20. 13 14 15 1 13 4 3 14 7 0

Page 17

1. 89 88 99 87 88 98 73 89
2. 888 898 659 977 887 858 959
3. 7,339 7,969 8,588 7,689 8,395 9,797
4. 649 89,559 67 794 982,488 78
5. 9,595 9,797 7,677 8,178 6,677 7,888
6. 9,959 7,389 8,554 6,929 9,879 5,998
7. 8,574 4,759 8,896 8,688 9,559 6,747
8. 39,279 84,578 75,958 47,768 89,925

Pages 19–20

1. 93 140 121 72 150 70
2. 131 92 100 153 101 162
3. 1,100 940 1,011 1,203 1,013 608
4. 813 1,486 412 339 702 1,001
5. 835 884 410 346 522 441
6. 648 541 801 593 330 475
7. 40 60 88 34 63 54
8. 67 92 203 113 206 141
9. 178 152 103 258 721 592
10. 329 471 651 580 1,002 1,104
11. 1,009 1,264 1,442 937 509 1,480
12. 266 197 231 1,752 1,358 873
13. 2,044 2,224 1,130 2,238 2,630 1,873
14. 245 236 378 279 211 248
15. 3,114 2,719 3,886 2,447 3,017
16. 16,334 9,458 18,806 18,113
17. 20,942 19,765 18,374 27,904
18. 10,154 13,346 15,879 15,233

Page 21

1. 182	137	$9.19
2. 963	839	$16.27
3. 727	900	$16.74
4. 4,746	5,166	$28.33
5. 84,442	82,307	$7.71
6. 16,334	9,458	$13.21

Page 22

1. 50	30	90
2. 80	60	70
3. 40	20	70

4. 70 + 5 = 75	18 + 50 = 68
5. 19 + 20 = 39	100 + 13 = 113
6. 40 + 12 = 52	60 + 21 = 81
7. 60 + 14 = 74	19 + 30 = 49
8. 30 + 20 + 9 = 59	20 + 30 + 6 = 56
9. 20 + 40 + 11 = 71	30 + 40 + 13 = 83
10. 15 + 20 + 30 = 65	21 + 20 + 50 = 91
11. 50 + 70 + 5 = 125	20 + 20 + 5 = 45

Pages 23–24

1. 300	100	3,100
650	4,580	12,490
2. 4,000	25,000	8,400
7,100	43,000	6,500
3. 50,000	76,000	220,000
9,000	70,000	2,000

4. 3,900 + 800 + 6,800 + 400 = 11,900
9,100 + 800 + 3,000 + 100 = 13,000

5. 800 + 3,300 + 200 + 5,400 = 9,700
500 + 2,200 + 700 + 5,800 = 9,200

6. 100 + 7,900 + 3,700 + 600 = 12,300
700 + 7,500 + 300 + 5,000 = 13,500

7. 76,000 + 7,000 + 27,000 = 110,000
20,000 + 6,000 + 28,000 = 54,000

8. 5,000 + 65,000 + 70,000 = 140,000
8,000 + 9,000 + 17,000 = 34,000

9. 28,000 + 93,000 + 3,000 = 124,000
42,000 + 2,000 + 19,000 = 63,000

10. 20 + 3,000 + 500 = 3,520
4,000 + 60 + 300 = 4,360

11. 100 + 4,000 + 80 = 4,180
200 + 8,000 + 10 = 8,210

12. 7,000 + 1,000 + 90 = 8,090
50 + 9,000 + 700 = 9,750

13. 600,000 + 40,000 + 6,000 = 646,000
100,000 + 1,000 + 500,000 = 601,000

14. 90,000 + 3,000 + 700,000 = 793,000
1,000 + 20,000 + 200,000 = 221,000

15. 6,000 + 200,000 + 5,000 = 211,000
5,000 + 800,000 + 5,000 = 810,000

Pages 25–28

1. 1,380 miles
2. 1,400 miles
3. $23,525
4. $24,000
5. $50,878
6. $51,000
7. 8,597 people
8. 8,600 people
9. $2,113.86
10. $2,120
11. 1,170 pounds
12. Yes
13. No
14. 915 calories
15. 930 calories
16. 1,386,308
17. 1,387,000
18. 1,380,000
19. 94,710 square miles
20. 95,000 square miles
21. 8,008,269
22. 7,900,000
23. $3,995
24. $4,000
25. 241 miles
26. 2002

Pages 29–30, Addition Review

1. 2,000
2. 14,800
3. $480,000
4. 7,469
5. 74,997
6. 599,887
7. 168
8. 295
9. 248
10. 1,143
11. 837
12. 1,190
13. 332
14. 542
15. 2,784
16. 3,923
17. 1,328,449
18. 981
19. 4,000 + 1,300 + 3,400 = 8,700
20. 70,000 + 71,000 + 34,000 = 175,000
21. 640,000
22. $789.41
23. 721 cars
24. **c.** 1,700 seats

Pages 31–32, Subtraction Skills Inventory

1. 44
2. 31
3. 301
4. 552
5. 18
6. 745
7. 7,086
8. 1,409
9. 44
10. 224
11. 1,052
12. 2,951
13. 489
14. 149
15. 16,984
16. 65,942
17. 46
18. 48
19. 7,500 − 2,400 = 5,100
20. 60,000 − 29,000 = 31,000
21. 490,000 − 300,000 = 190,000
22. $462.67
23. $13.05
24. 1,134 miles
25. 1,136 voters

Pages 33–34

1. 2 1 5 8 8 5 8 2 3 0
2. 6 3 6 8 1 1 9 1 6 1
3. 2 5 8 9 3 9 3 0 6 3
4. 0 9 3 1 3 7 5 4 0 5
5. 7 3 7 0 5 7 7 0 4 4
6. 9 5 6 6 2 4 4 6 1 7
7. 9 2 5 1 4 4 4 6 9 7
8. 4 6 5 8 0 9 2 8 6 3
9. 2 0 7 7 7 9 0 4 2 8
10. 1 8 1 5 8 3 9 2 2 7
11. 5 5 3 3 1 1 2 9 3 3
12. 3 7 0 3 5 0 9 6 4 7
13. 5 4 9 6 0 8 2 7 0 8
14. 1 3 2 1 8 2 6 8 9 1
15. 8 9 6 9 3 4 7 0 7 4
16. 6 4 1 2 4 6 4 8 2 3
17. 7 9 2 8 8 2 2 8 8 0
18. 6 1 6 5 3 0 0 1 5 5
19. 7 7 4 5 2 6 9 1 4 7
20. 5 9 6 0 7 4 1 5 9 7

Page 35

1. 12 34 7 20 24 41 51 21
2. 20 14 4 34 26 20 22 41
3. 115 310 413 201 161 215 406
4. 535 300 481 331 611 36 401
5. 4,351 1,154 4,130 174 6,325 1,133
6. 122 1,206 1,205 3,607 3,633 4,770
7. 54,122 2,508 70,322 52,305 18,114
8. 530,133 14,240 500,732 243,382 663,152

Page 36

1. 73 19 14 87 58 85 53 29

2. 58 56 22 19 39 25 29 38

3. 213 715 909 328 707 613 317

4. 102 137 15 118 118 244 148

5. 109 464 318 316 413 119 219

Pages 37–38

1. 894 61 481 481 466 242

2. 278 89 58 78 174 398

3. 2,839 3,489 1,989 4,999 6,087 909

4. 2,963 4,719 208 2,409 2,803 4,209

5. 839 2,784 2,996 4,897 2,942 2,691

6. 1,887 2,948 976 780 6,785 888

7. 32,871 29,691 77,891 68,571 28,971

8. 83,351 64,788 80,584 40,899 79,969

9. 83,988 78,248 35,897 26,899 58,887

10. 50,989 9,597 20,058 21,789 23,909

11. 2,899 5,688 11,857 8,888 8,989

12. 580,707 268,637 144,784 680,647 868,123

13. 74,287 185,489 190,888 375,971 298,907

14. 125,437 222,155 1,585 148,492 187,609

Page 39

1. 6,551 8,503 5,635

2. 89 2,355 6,809

3. 789 1,083 778

4. 47,891 19,095 79,361

5. 17,887 20,868 49,127

6. 18,997 20,839 759

Pages 40–41

1. 239 159 278 27 8 116

2. 537 173 267 109 218 6

3. 2,031 4,581 1,660 3,382 8,030 5,571

4. 7,374 8,679 3,449 2,567 7,708 699

5. 1,439 2,428 439 6,088 69 6,776

6. 4,564 3,772 7,073 467 5,593 6,684

7. 6,456 5,367 571 1,662 1,574 1,742

8. 503 1,752 2,570 3,624 1,575 734

9. 1,046 14,636 18,544 26,188 27,693

10. 16,443 28,718 40,567 461 26,127

Page 42

1. 309 547 $8.73

2. 772 427 $2.56

3. 444 179 $2.91

4. 3,587 5,494 $17.62

5. 1,103 3,594 $39.49

6. 7,566 3,076 $83.97

7. 17,946 56,882 $335.83

Page 43

1. 32 37 43

2. 28 11 66

3. 45 34 69

4. $87 - 30 = 57$ $106 - 90 = 16$ $53 - 20 = 33$

5. $97 - 50 = 47$ $76 - 40 = 36$ $115 - 80 = 35$

6. $97 - 70 = 27$ $89 - 60 = 29$ $66 - 30 = 36$

7. $98 - 40 = 58$ $88 - 20 = 68$ $134 - 50 = 84$

Page 44

1. 42,400 – 9,500 = 32,900
2. 38,200 – 8,500 = 29,700
3. 87,400 – 9,500 = 77,900
4. 78,300 – 9,700 = 68,600
5. 34,800 – 5,800 = 29,000
6. 92,300 – 9,000 = 83,300
7. 22,000 – 20,000 = 2,000
8. 63,000 – 57,000 = 6,000
9. 55,000 – 43,000 = 12,000
10. 26,000 – 17,000 = 9,000
11. 678,000 – 97,000 = 581,000
12. 328,000 – 59,000 = 269,000
13. 210,000 – 120,000 = 90,000
14. 520,000 – 420,000 = 100,000
15. 810,000 – 330,000 = 480,000
16. 470,000 – 210,000 = 260,000
17. 400,000 – 110,000 = 290,000
18. 190,000 – 100,000 = 90,000
19. 2,500,000 – 400,000 = 2,100,000
20. 6,500,000 – 1,200,000 = 5,300,000
21. 9,100,000 – 1,300,000 = 7,800,000
22. 2,700,000 – 2,600,000 = 100,000

Pages 45–48

1. 105°
2. 100°
3. $21.55
4. $20
5. 65 hits
6. 70 hits
7. $20,582
8. $20,000
9. 190 miles
10. 190 miles
11. $114,325
12. $115,000
13. 787 apartments
14. 800 apartments
15. $62,510
16. $63,000
17. $6,214
18. $6,000
19. 228 years
20. $91.20
21. 1,588 miles
22. 19 years old
23. 7 feet
24. 785 feet
25. 63,271,000 people
26. 63,300,000 people
27. 64,000,000 people

Pages 49–50

1. 8 seats
2. $12,219
3. 97
4. 335
5. $688
6. $208
7. $210
8. 719,354 newspapers
9. 5,306,000 newspapers
10. $1,180
11. 2,256,827 cars
12. 6,814,000 cars

Pages 51–52, Subtraction Review

1. 800,000
2. 250,000
3. 421
4. 766
5. 588
6. 249
7. 2,211
8. 308
9. 4,435
10. 15,992
11. 85,500
12. 4,740
13. 248
14. 8,759
15. 196,487
16. 55,844
17. 31,977

18. 348,000 − 198,000 = 150,000
19. 890,000 − 60,000 = 830,000
20. 2,500,000 − 1,800,000 = 700,000
21. 94,765 permanent residents
22. $6,425
23. $484.50
24. 950,000 trucks
25. 269,000 trucks

Pages 54–55, Multiplication Skills Inventory

1. 126
2. 305
3. 2,169
4. 32,408
5. 1,196
6. 56,420
7. 17,408
8. 263,246
9. 336
10. 2,124
11. 3,108
12. 4,002
13. 160,272
14. 519,082
15. 4,331,820
16. 46,575,782
17. 30,045,600
18. 4,700
19. 526,000
20. 1,377,144
21. 3,238,488
22. 1,440 miles
23. $30,576
24. 3,717 miles
25. 1,960 miles

Pages 56–58

1. 32 11 56 21 0
2. 60 18 84 16 22
3. 20 54 30 48 44
4. 12 132 0 49 12
5. 24 64 12 18 40
6. 45 120 121 48 30
7. 110 81 36 72 33
8. 42 16 32 132 27
9. 63 7 99 96 55
10. 42 5 24 24 36
11. 72 25 36 48 88
12. 56 28 72 77 100
13. 42 81 45 110 64
14. 121 18 12 40 36
15. 24 0 120 49 12
16. 30 132 30 12 48
17. 22 20 18 44 54
18. 84 21 16 0 32
19. 11 60 56 63 21
20. 54 144 66 45 72
21. 35 27 100 108 77
22. 88 56 25 28 36
23. 24 48 42 72 36
24. 5 99 24 96 27
25. 55 63 16 6 32
26. 72 132 33 48 72
27. 54 45 35 108 21
28. 27 144 63 28 66
29. 108 72 25 42 27 72 45
30. 12 12 22 21 66 100 48
31. 24 7 42 121 120 12 54
32. 60 21 77 36 5 63 33
33. 110 24 132 20 0 27 54
34. 88 24 99 16 48 64 0
35. 30 18 32 144 45 56 36
36. 96 81 18 49 48 84 11
37. 63 35 72 55 28 32 132
38. 36 40 30 44 16 56 108

Page 60

1. 639 567 248 129 728 270 160 490
2. 1,248 2,463 5,499 2,406 2,008 5,409 5,607
3. 4,270 1,260 4,080 990 3,500 6,300 1,200
4. 16,804 9,306 10,042 48,088 42,606 56,077

Pages 61–62

1. 1,426 713 1,767 3,956 528 1,768 6,075
2. 1,848 6,319 3,213 903 2,952 6,006 3,538
3. 3,650 870 5,760 3,640 360 3,760 1,440
4. 50,456 34,680 454,879 310,114 306,246 7,098,078
5. 3,280 5,680 1,560 4,550 2,480 1,290 2,790
6. 12,820 12,480 35,770 14,260 27,360 32,880
7. 6,607,473 39,765,467 18,875,168 1,257,060 60,785,416
8. 5,759,376 9,327,096 7,421,841 3,724,264 13,826,916

Pages 63–64

1. 123 405 128 159
2. 1,482 969 3,055 1,684
3. 7,613 25,502 27,608 20,086
4. 23,199 77,697 22,838 46,360
5. 6,996 1,333 1,606 12,369
6. 18,820 21,390 64,880 36,480
7. 11,340 21,080 60,750 29,760
8. 301,323 75,262 128,604 163,829
9. 1,638,752 1,685,088 1,897,231
10. 80,840 2,517,262 3,111,051
11. 4,800,600 846,741 518,676

Pages 65–67

1. 68 320 665 152 747 258 198 416
2. 170 170 344 126 81 372 324 432
3. 138 324 144 255 130 336 296 424
4. 265 132 252 132 675 651 344 376
5. 516 256 231 232 315 165 194 468
6. 414 208 252 275 511 414 296 36
7. 291 387 304 198 304 342 91 140
8. 185 432 138 657 372 434 54 510
9. 456 76 208 235 76 477 192 651
10. 294 330 522 584 255 56 156 78
11. 1,527 2,442 4,015 5,418 5,664 218 2,842
12. 1,185 792 5,068 1,158 6,600 3,092 2,058
13. 7,758 2,262 4,902 2,815 1,436 1,340 6,642
14. 972 4,756 2,795 2,726
15. 5,481 2,516 3,604 2,688
16. 1,534 3,410 3,182 1,638
17. 4,648 5,472 3,192 6,138
18. 1,971 1,856 4,950 8,184
19. 10,965 20,552 49,217 29,792
20. 60,860 26,208 21,624 24,489
21. $140.30 $537.56 $587.65 $169.44
22. $5,778.99 $1,138.86 $4,058.64 $3,160.17
23. $1,029.92 $2,077.04 $4,898.88 $2,520.10

Pages 68–69

1. 470 530 920 360 710 280 650
2. 3,680 8,660 7,610 9,460 4,790 2,610
3. 948,300 235,600 307,900 430,800 557,000 609,000
4. 7,400 800 25,600 3,100 20,900 6,800
5. 73,000 421,000 16,000 208,000 450,000 623,000
6. 740 310 250 560
7. 980 460 500 370
8. 6,620 2,960 8,020 4,290
9. 5,060 3,800 4,090 1,100
10. 1,280 8,390 7,560 2,170
11. 65,780 73,080 88,150 70,490
12. 2,600 4,300 9,400 3,700
13. 9,500 3,000 6,300 7,400
14. 8,100 5,700 6,800 2,500
15. 95,700 21,400 69,300 89,800
16. 58,100 14,200 81,300 42,000
17. 22,500 73,000 20,700 39,600
18. 65,000 88,000 62,000 91,000
19. 32,000 603,000 540,000 931,000
20. 46,000 261,000 380,000 715,000
21. 425,000 689,000 237,000 499,000

Pages 70–72

1. 320 54,000 36,000
2. 630 3,500 54,000
3. 300 16,000 26,000
4. 1,400 2,000 560,000
5. $4 \times 800 = 3{,}200$ $7(300) = 2{,}100$ $3(400) = 1{,}200$
6. $(900)(3) = 2{,}700$ $600 \times 2 = 1{,}200$ $8 \cdot 900 = 7{,}200$
7. $6 \cdot 500 = 3{,}000$ $(5)(900) = 4{,}500$ $200 \times 4 = 800$

8. $2(4{,}000) = 8{,}000$ $(8{,}000)(4) = 32{,}000$
$4(6{,}000) = 24{,}000$

9. $7 \times 3{,}000 = 21{,}000$ $5 \cdot 4{,}000 = 20{,}000$
$8 \cdot 1{,}000 = 8{,}000$

10. $3 \cdot 6{,}000 = 18{,}000$ $9{,}000 \times 6 = 54{,}000$
$(2)(18{,}000) = 36{,}000$

11. $30 \cdot 70 = 2{,}100$ $60(200) = 12{,}000$
$(80)(4{,}000) = 320{,}000$

12. $10 \times 300 = 3{,}000$ $90 \times 30 = 2{,}700$
$200 \times 700 = 140{,}000$

13. $5{,}000 \cdot 70 = 350{,}000$ $700 \cdot 20 = 14{,}000$
$400 \cdot 1{,}000 = 400{,}000$

14. **b.** 7,614
15. **c.** 74,808
16. **a.** 4,088
17. **d.** 243,126
18. **c.** 15,300
19. **a.** 132,210
20. **d.** 8,736
21. **b.** 361,368
22. **d.** 46,224
23. **d.** 321,432
24. **d.** 20,064
25. **b.** 524,286

Pages 73–76

1. 212 miles
2. 200 miles
3. 204 miles
4. 210 miles
5. 2,494 seats
6. 2,400 seats
7. $559.96
8. $560
9. 1,550 words
10. 1,800 words
11. $7,506
12. $8,000
13. 324 miles
14. 300 miles
15. $3,597,200
16. $3,500,000
17. $40.81
18. $42
19. $1,200
20. $3,614
21. $3,000
22. 627 students
23. 600 students
24. 564 miles
25. 500 miles
26. 5,500 feet

Pages 77–78

1. $737.30
2. 576 cans
3. $129,600
4. $1,037
5. 1,344 apartments
6. **c.** 5,200
7. $23,040
8. $2,522
9. 247 miles
10. $307
11. $15,188
12. $360
13. $108
14. 1,560 containers

Pages 79–80, Multiplication Review

1. $9,500
2. 9,000 or $9,000
3. 57
4. 105,273
5. 11,314
6. 248,763
7. 248
8. 1,263
9. 35,380
10. 14,796
11. 438
12. 3,656
13. 4,165
14. 39,008
15. 3,903,632
16. 53,000
17. 42,300
18. 72,400
19. $(400)(800) = 320{,}000$
20. $(500)(7{,}000) = 3{,}500{,}000$
21. $(900)(50{,}000) = 45{,}000{,}000$
22. **d.** $6,000
23. 23,000 meters
24. 2,555 hours
25. 125 words per minute

Pages 82–83, Division Skills Inventory

1. 23
2. 67
3. 238
4. 418 r 5
5. 704 r 1
6. 4,802 r 4
7. 7
8. 5 r 21
9. 9
10. 83
11. 91 r 20
12. 624 r 9
13. 58
14. 70 r 27
15. 70
16. $2{,}800 \div 7 = 400$
17. $18{,}000 \div 6 = 3{,}000$
18. **c.** 580
19. 22 hours
20. 163 pounds
21. $656
22. $39.80

Pages 84–86

1. 3 9 5
2. 7 3 6
3. 4 9 3
4. 8 8 6
5. 9 9 8
6. 4 7 3
7. 9 7 7
8. 6 6 3
9. 8 8 6
10. 5 8 8
11. 9 6 4
12. 9 4 7
13. 9 10 8
14. 2 11 4
15. 5 5 6
16. 5 0 5
17. 4 7 12
18. 5 10 2
19. 12 3 7
20. 2 6 5
21. 12 8 3
22. 8 12 2
23. 4 2 6
24. 2 4 7
25. 2 7 9
26. 9 0 11
27. 4 3 4
28. 3 11 10
29. 5 10 8
30. 7 8 0
31. 4 11 9
32. 2 6 4
33. 12 12 10
34. 4 3 11
35. 2 11 11
36. 3 2 6
37. 9 10 6
38. 10 7 12
39. 12 7 5

Pages 88–89

1. 42 89 53 46 79
2. 54 89 63 84 86
3. 43 68 65 67 97
4. 34 72 66 48 29
5. 68 48 59 87 60
6. 90 60 40 70 90
7. 346 293 388 452
8. 775 653 413 716
9. 604 307 904 608
10. 709 431 617 572
11. $3.80 $6.61 $3.12 $5.08
12. $6.56 $2.91 $2.03 $5.36
13. $7.40 $2.21 $11.85 $9.86

Pages 90–92

1. 241 r 3 703 r 5 470 r 2 294 r 1
2. 607 r 3 415 r 5 512 r 1 910 r 7
3. 809 r 1 368 r 2 420 r 6 716 r 3
4. 536 r 7 708 r 3 661 r 5 419 r 2
5. 188 r 5 269 r 4 432 r 3 751 r 6
6. 180 r 3 438 r 4 307 r 4 880 r 2
7. 2,354 r 2 3,913 r 1 5,226 r 2 8,626 r 4
8. 8,401 r 3 4,067 r 2 8,240 r 3 7,117 r 1
9. 2,083 r 3 4,059 r 4 6,502 r 6 2,083 r 5
10. 5,004 r 2 3,107 r 6 7,028 r 1 4,800 r 3
11. 9,120 r 5 6,080 r 7 5,006 r 5 7,102 r 2
12. 614 r 4 9,774 r 5 27,737 r 1 1,908 r 4
13. 2,726 r 7 19,633 r 2 15,547 r 1 6,425 r 6
14. 8,713 r 7 1,611 r 5 8,750 r 4 13,913 r 3
15. 4,096 r 2 3,907 r 3 9,560 r 2 2,814 r 5

Page 93

1. 8
2. 7
3. 4
4. 9
5. 7
6. 4
7. 8
8. 9
9. 0
10. 0
11. 1
12. 5
13. 0
14. 7
15. 1
16. 7
17. 8
18. 0
19. 8
20. 0

Pages 95–97

1. 8 7 5 4
2. 6 8 4 9
3. 7 6 4 3
4. 2 5 6 7
5. 8 3 4 6
6. 5 9 3 10
7. 7 6 4 7
8. 8 5 9 6
9. 8 r 27 6 r 25 4 r 6 3 r 15
10. 4 r 20 3 r 12 6 r 17 7 r 18
11. 5 r 31 4 r 5 3 r 50 8 r 6
12. 3 r 23 5 r 16 4 r 22 9 r 12
13. 23 45 62 81
14. 92 63 42 71
15. 24 36 23 42
16. 40 r 8 30 r 7 50 r 9 60 r 2
17. 52 r 8 41 r 10 53 r 12 11 r 17
18. 234 521 633 418
19. 536 823 455 629
20. 408 507 309 608
21. 540 320 760 490

Pages 98–99

1. 24 38 51 63
2. 18 r 506 22 r 421 46 r 210 34 r 38
3. 56 72 83 29
4. 36 91 87 53
5. 40 r 19 60 r 32 70 r 43 50 r 56
6. 72 r 3 88 r 12 36 r 15 63 r 40
7. 52 88 67 27

Pages 100–101

1. 300 200 300
2. 400 700 10,000
3. 80 1,100 800
4. 90 3,000 5,000
5. 6 90 500
6. 90 60 10
7. 4 70 50
8. 300 8 800
9. $420 \div 7 = 60$ $400 \div 4 = 100$
10. $300 \div 6 = 50$ $490 \div 7 = 70$
11. $200 \div 5 = 40$ $960 \div 8 = 120$
12. $210 \div 3 = 70$ $720 \div 9 = 80$

13. 3,200 ÷ 4 = 800 4,000 ÷ 8 = 500
14. 3,000 ÷ 6 = 500 2,100 ÷ 7 = 300
15. 8,100 ÷ 9 = 900 3,000 ÷ 5 = 600
16. 1,800 ÷ 3 = 600 4,800 ÷ 6 = 800

Pages 102–103

1. 130 240 300 430
2. 720 500 350 700
3. 1,200 2,400 3,500 6,100

4. d. 283	**8. a.** 833	**12. d.** 3,903
5. b. 721	**9. d.** 298	**13. a.** 5,471
6. c. 428	**10. b.** 2,667	**14. b.** 2,918
7. c. 307	**11. c.** 4,458	**15. c.** 1,784

Pages 104–108

1. $29.75	**11.** $574.50	**21.** 8 cubic yards
2. $30	**12.** 14 hours	**22.** 6 gallons
3. 21 points	**13.** $436	**23.** 60 mph
4. 20 points	**14.** 384 miles	**24.** 510 gallons
5. 44 rows	**15.** 287 containers	**25.** 26 months
6. 40 rows	**16.** 13 inches	**26.** 21 shelves
7. $39.80	**17.** $2,400	**27.** 26 crates
8. $40	**18. c.** 40 minutes	**28.** 28 cartons
9. $5.99	**19.** 5 inches	**29.** 9 months
10. $6	**20.** 12 months	**30.** 33 months

Pages 109–110

1. 35 hours	**8.** 48 people
2. $2,900	**9.** 445 miles
3. 5 hours	**10.** 140 months
4. $58.20	**11.** $6,700
5. $1,860.50	**12.** $35,500
6. 84	**13.** $9,194
7. $3	**14.** $7,519

Pages 111–113, Division Review

1. 14,000	**4.** 36,041	**7.** 5,539	**17.** 681
2. 800	**5.** 456	**8.** 166,944	**18.** 72
3. 5,043	**6.** 6,111	**9.** 318	**19.** 210,000
		10. 2,088	**20.** 8,000
		11. 9,288	**21.** 2,905 miles
		12. 724,000	**22.** 13 cans
		13. 43	**23.** $770
		14. 627 r 6	**24.** 23 miles per gallon
		15. 8	**25.** $2,100
		16. 72	

Pages 114–118, Posttest A

1. 6	**13.** 142	**25.** 126,728
2. 300	**14.** 3,246	**26.** 150,000
3. 290,000	**15.** 665	**27.** 175 miles
4. $3,700	**16.** 221,860	**28.** $2,940
5. 669	**17.** 78,670	**29.** 78
6. 22,435	**18.** 33,000	**30.** 904 r 5
7. 170	**19.** $8,569	**31.** 83
8. 209,627	**20.** $16,035	**32.** 460
9. 77,893	**21.** 736	**33.** 74
10. 23,800	**22.** 13,888	**34.** 5,000
11. 1,075 miles	**23.** 5,963	**35.** 46 rows
12. $30,700	**24.** 382,000	**36.** 219 boxes

Pages 120–121

1. 12 pounds	**13.** 40 yd
2. 26,400 feet	**14.** 6 buckets
3. 4 kilometers	**15.** 200 cm
4. 18 yards	**16.** 324 cu ft
5. 288 quarts	**17.** 20 sq yd
6. 12,000 grams	**18.** 3 lb
7. 14 yards	**19.** 6,000 lb
8. 480 minutes	**20.** 4 gal
9. 42,500 centimeters	**21.** 72 in.
10. 74 quarts	**22.** 5 cu yd
11. 9 miles	**23.** 18 sq ft
12. 105 gallons	

Page 122

1. 9 hr 25 min
2. 8 gal 2 qt
3. 12 lb 7 oz
4. 15 yd
5. 12 days 10 hr
6. 6 m 20 cm
7. 15 ft
8. 14 kg 465 g
9. 25 min 54 sec
10. 13 T 1,525 lb
11. 17 wk 1 day
12. 9 days 13 hr

Page 123

1. 3 ft 9 in.
2. 1 day 17 hr
3. 2 lb 10 oz
4. 2 gal 2 qt
5. 3 m 53 cm
6. 5 min 32 sec
7. 2 hr 25 min
8. 1,240 lb
9. 2 km 695 m
10. 2 yd 1 ft
11. 64 cm
12. 4 wk 5 days

Page 124

1. 10 ft 8 in.
2. 19 gal 2 qt
3. 43 lb 12 oz
4. 61 T 1,500 lb
5. 33 m 44 cm
6. 26 wk 4 days
7. 20 yd
8. 43 km 200 m
9. 22 hr 15 min
10. 37 kg 120 g
11. 130 yd 2 ft
12. 34 days 12 hr

Page 125

1. 1 lb 6 oz
2. 2 yd 1 ft
3. 3 gal 2 qt
4. 1 T 700 lb
5. 2 km 300 m
6. 3 ft 11 in.
7. 3 lb 14 oz
8. 2 wk 4 days
9. 2 m 18 cm
10. 2 min 8 sec
11. 2 kg 355 g
12. 2 hr 15 min

Page 127

1. 82 feet
2. $105.78
3. 176 feet
4. $68.64
5. 23 feet
6. 120 inches
7. 10 feet
8. 14 feet

Pages 128–130

1. 216 square feet
2. 24 square yards
3. $407.76
4. $482.76
5. 1,512 tiles
6. $1,179.36
7. $1,299.16
8. 25 feet
9. 7 gallons
10. $139.65
11. 112 feet
12. 391 square feet
13. 864 square inches
14. 378 square feet

Pages 131–132

1. 768 cubic feet
2. 810 cubic inches
3. 27 cubic inches
4. 135 ice cubes
5. 72 cubic feet
6. 8,640 cubic feet
7. 480 cartons
8. 30 truckloads
9. $750
10. 48 boxes

Page 133

1. $9.55
2. $4.00
3. $31.65
4. $25.15
5. $32.10

Pages 134–135

1. **a.** 9¢ **b.** 5¢ **c.** 22¢ **d.** 25¢ **e.** 32¢ **f.** 13¢ **g.** 19¢ **h.** 26¢ **i.** 30¢ **j.** 35¢
2. $0.57
3. $10.12
4. $4.88
5. $0.24
6. $1.90
7. $39.55
8. $1.51
9. $3.34
10. $1.93
11. $5.97

Pages 136–137

1. $6.35
2. $1.50
3. $2.19
4. $0.20
5. $5.22
6. $0.25
7. $42.26
8. $3.01
9. $6.88
10. $2.00
11. $10.76
12. $1.01

Pages 138–139

1. 580 calories
2. 1,085 calories
3. 1,010 calories
4. 1,300 calories
5. 325 calories
6. 1,485 calories
7. 1,000 calories
8. 1,000 calories
9. 930 calories
10. 730 calories

Pages 140–141

1. $498.34
2. $353.83
3. $1,973.40
4. $548.00
5. $438.70
6. $10,963.48
7. $25,913.68
8. $18,399.16
9. $10,614.90
10. $4,664.40
11. $188.83
12. $81.33
13. $42.73
14. **d.** $175

Pages 142–143

1. 50¢
2. $0.90
3. $1.05
4. $1.60
5. $1.10
6. $2.10
7. $0.85
8. $1.25
9. $1.40
10. $1.95
11. $136.05
12. $149.18
13. $104.67
14. $363.35
15. $118.03
16. $122.49
17. $205.35

Pages 144–145

1. $29,285.00
2. $1,698.53
3. $5,271.32
4. $1,171.42
5. $409.99
6. $8,551.26
7. $20,733.74
8. $1,730
9. $6,871.20
10. $13,862.54
11. $6,136
12. $7,726.54
13. $498.00
14. $7,228.54

Pages 146–147

1. 9¢
2. $2.70
3. $6.30
4. 40¢
5. 12¢
6. $3.60
7. $1.44
8. $43.20
9. 25¢
10. $11.70

1.80	0.50
9.00	3.60
0.15	3.90
0.20	$30.85

Page 149

1. 11:50 A.M.
2. 12:42 P.M.
3. 52 minutes
4. 53 minutes
5. Yes
6. No, not on Saturday or Sunday
7. 53 minutes
8. $187.50
9. $375
10. 6 hr 38 min
11. $286
12. $638
13. $352

Pages 150–151

1. $1,440
2. $294
3. $2,315
4. $660
5. $7,440
6. $1,900
7. $168
8. $19
9. $27
10. 60 baseball caps
11. 62 T-shirts
12. $39
13. 250 bags
14. $12,070

Pages 152–153

1. 1,744 miles
2. 11 miles
3. 352 miles
4. 10 miles
5. 402 miles
6. 948 feet
7. 414 miles
8. 15 hours
9. 29 mph
10. 2 hours
11. 1,420 mph
12. 64 mph
13. 50 mph

Page 154

1. $119.85
2. $211.95
3. $174.75
4. $110.85
5. $176.40
6. $307.75

Pages 155–158, Posttest B

1. **a.** 63,399
2. **b.** 5,274
3. **c.** 800
4. **d.** $648
5. **d.** $54
6. **b.** 2
7. **c.** 607
8. **b.** 195,741
9. **c.** 4,383
10. **b.** 110
11. **d.** 674
12. **c.** $8,124
13. **b.** 17
14. **d.** 305
15. **a.** 2,936
16. **a.** 26,492
17. **c.** 46,485
18. **d.** 64,624
19. **d.** $60,000
20. **d.** 1,332
21. **b.** 333
22. **d.** 168
23. **a.** 41,400
24. **c.** 410
25. **c.** 26,000
26. **c.** 250,000
27. **c.** 293,000
28. **d.** 2,661
29. **b.** 4,000
30. **b.** $251.12
31. **a.** $354.44
32. **d.** 3,800
33. **c.** 7,000,000
34. **d.** $100
35. **b.** $280

UNITS OF MEASUREMENT

The table below lists common units of measure and what they are equal to in other units. The terms *meter* and *kilogram* are part of the **metric system** of measurement. The metric system is used in many countries and more frequently now in the United States. However, the **U.S. customary system** of measurement is most often used here.

Common abbreviations for each unit of measurement are given in parentheses.

Distance	
1 foot (ft or ')	= 12 inches (in. or ")
1 yard (yd)	= 3 feet
1 mile (mi)	= 5,280 feet
1 meter (m)	= 1,000 millimeters (mm)
1 meter	= 100 centimeters (cm)
1 kilometer (km)	= 1,000 meters

Area	
1 square foot (sq ft)	= 144 square inches (sq in.)
1 square yard (sq yd)	= 9 square feet
1 acre	= 43,560 square feet

Volume	
1 cup (c)	= 8 fluid ounces (fl oz)
1 quart (qt)	= 4 cups
1 gallon (gal)	= 4 quarts
1 gallon	= 231 cubic inches (cu in.)
1 cubic foot (cu ft)	= 1,728 cubic inches
1 cubic yard (cu yd)	= 27 cubic feet
1 board foot	= 1 inch by 3 inches by 23 inches

Weight/Mass	
1 pound (lb)	= 16 ounces (oz)
1 ton (T)	= 2,000 pounds
1 kilogram (kg)	= 1,000 grams (gr)

Time	
1 minute (min)	= 60 seconds (sec)
1 hour (hr)	= 60 minutes
1 day	= 24 hours
1 week (wk)	= 7 days
1 year (yr)	= 365 days

GLOSSARY

A

addend One of the numbers in an addition problem. In $3 + 2 = 5$ the addends are 3 and 2.

addition The mathematical operation used to find a sum. The problem $8 + 1 = 9$ is an example.

approximate Another word for *estimate.* As an adjective it means "close or almost exact." A jacket costs $79. The approximate price is $80. The symbol $\approx$ means "approximately equal to."

area A measure of the amount of surface on a flat figure. A room that is 12 feet long and 10 feet wide has an area of $12 \times 10 = 120$ square feet.

average A sum divided by the number of items that make up the sum. An average, also called the *mean,* is a representative number for a group. If the low temperature one day was 24° and the low temperature the next day was 28°, the average low temperature for the two days is the sum ($24° + 28° = 52°$) divided by 2 (the number of days). $52° \div 2 = 26°$

B

borrowing Regrouping the digits in the top number of a subtraction problem. In the problem $84 - 29$, the 8 in the tens column becomes 7, and the 4 in the units column becomes 14.

C

carrying Regrouping the digits in an addition or multiplication problem. In the problem $37 + 9$, the sum of the units column is $7 + 9 = 16$. The digit 6 remains in the units column, and the digit 1 is added to the tens column.

commutative property for addition Numbers can be added in any order. The sum of $4 + 9$ is the same as the sum of $9 + 4$.

commutative property for multiplication Numbers can be multiplied in any order. The product of 3×6 is the same as the product of 6×3.

compatible pairs Two numbers with which it is easy to perform a mathematical operation. The numbers 83 and 7 are a compatible pair. They are easy to add because their sum ends with zero.

cubic units A category of measurement for volume. The volume of a tank is usually measured in cubic inches or cubic feet.

D

difference The answer to a subtraction problem. In the problem $14 - 9 = 5$, the difference is 5.

digit One of the ten number symbols. The digits are 0, 1, 2, 3, 4, 5, 6, 7, 8, and 9.

distance A measure of the length of a straight line between two points. The distance between Chicago and Milwaukee is 90 miles.

dividend The number in a division problem into which another number divides. In $21 \div 7 = 3$, the dividend is 21.

division A mathematical operation that requires figuring out how many times one amount is contained in another. In the problem $20 \div 5 = 4$, the answer means that there are exactly four fives in twenty.

divisor The number in a division problem that divides into another. In $28 \div 7 = 4$, the divisor is 7.

E

estimate As a noun: an approximate value. The population of a village is 1,836. The number 1,800 is an estimate of the population. As a verb: to find an approximate value. You may be asked to estimate the number of people living on your street.

F

factor A number that, when multiplied by another number, results in a product. The numbers 2 and 3 are both factors of 6.

formula A mathematical rule written with an = sign. The formula for calculating distance is $d = rt$, where d is distance, r is rate (usually in miles per hour), and t is time.

front-end rounding Rounding the left-most digit of each number in a problem in order to calculate an estimate. In the problem 87×638, the number 87 rounds to 90, and the number 638 rounds to 600. The estimate is $90 \times 600 = 54{,}000$.

H

height The straight-line measurement from the base of an object to the top. The measurements of a rectangular container include the length, the width, and the height.

K

kilogram The standard unit of weight in the metric system. A kilogram is a little more than 2 pounds.

L

label A word or abbreviation used to identify the unit of measurement of some quantity. A crate has a weight of 14 pounds. The label is *pounds.*

length A straight-line measurement. The longer side of a rectangle is usually called the length.

M

mean Another word for *average.* A sum divided by the number of items that make up the sum.

measurement A dimension, quantity, or capacity. The measurements of a room usually include the length, the width, and the height.

meter The standard unit of length in the metric system. A meter is a little more than 1 yard.

metric system A standard of measure based on tens, hundreds, and thousands. The standard unit of length in the metric system is the meter. The standard unit of weight is the kilogram. The standard unit of liquid measure is the liter.

minuend The number in a subtraction problem from which another number is subtracted. In the problem $9 - 2 = 7$, the minuend is 9.

minus Reduced by some amount. For example, seven minus one is six. The minus sign (−) is the symbol used for subtraction: $7 - 1 = 6$.

multiplicand The number in a multiplication problem that is multiplied by another number. In the problem $25 \times 4 = 100$, the multiplicand is 25.

multiplication A mathematical operation with whole numbers that consists of adding a number (the multiplicand) a certain number of times. For example, the problem $6 \times 3 = 18$ means finding the sum of three sixes: $6 + 6 + 6 = 18$.

multiplier The number in a multiplication problem by which another number is multiplied. In the problem $12 \times 4 = 48$, the multiplier is 4.

O

operation A mathematical process such as addition, subtraction, multiplication, or division.

P

parallel Being an equal distance apart. The sides opposite each other in a rectangle are parallel.

partial product In a multiplication problem, the result of multiplying the multiplicand by one digit of the multiplier. In the problem 125×13, the partial product of multiplying 125 by 3 is 375.

perimeter A measure of the distance around a flat figure. A rectangle that is 3 feet long and 2 feet wide has a perimeter 10 feet. $3 + 3 + 2 + 2 = 10$

place value The number that every digit stands for. For example, in 235, the digit 3 stands for 30 because 3 is in the tens place. The digit 2 stands for 200 because 2 is in the hundreds place.

plus Increased by some amount. For example, two plus seven is nine. The plus sign (+) is the symbol used for addition: $2 + 7 = 9$.

product The answer to a multiplication problem. In the problem $3 \cdot 6 = 18$, the product is 18.

Q

quotient The answer to a division problem. In the problem $63 \div 9 = 7$, the quotient is 7.

R

rate An amount whose unit of measure contains a word such as *per* or *for each.* For example, the speed (or rate of speed) of a moving vehicle is often measured in miles per hour.

rectangle A four-sided flat shape with four square corners. A page of a newspaper is a rectangle.

remainder A number left over when one number divides into another. The answer to $25 \div 6$ is 4 with a remainder of 1.

rounding Making an estimate, ending in zeros, that is close to an original value. 83 rounded to the nearest ten is 80. 1,924 rounded to the nearest thousand is 2,000.

S

short division Mentally multiplying and subtracting in a division problem. This division method works best with one-digit divisors. For example, $1{,}302 \div 6$ becomes:

$$\begin{array}{r} 2\ 1\ 7 \\ 6\overline{)1{,}3^{1}0^{4}2} \end{array}$$

square A four-sided flat figure with four right angles and four equal sides

square units A category of measurement for area. The area of a floor is measured in square feet or square yards or square meters.

subtraction The mathematical operation used to find the difference between two numbers. The problem $14 - 5 = 9$ is an example.

subtrahend The number in a subtraction problem that is subtracted from another. In the problem $8 - 5 = 3$, the subtrahend is 5.

sum The answer to an addition problem. In the problem $4 + 9 = 13$, the sum is 13.

symbol A printed sign that represents an operation of a quantity or a relation. The symbol + means "add." The symbol ° means "degree." The symbol = means "is equal to."

T

table An orderly arrangement of numbers in rows and columns. Train schedules are often in the form of a table.

time A measurement from a point in the past to a more recent point. The units of measurement for time are seconds, minutes, hours, days, weeks, months, years, and so on.

total Another word for *sum;* the answer to an addition problem

U

units The name of the right-most place in the whole number system. In the number 623, the digit 3 is in the units place.

units of measurement Labels for determining a quantity. For example, pounds and ounces are units of measurement for weight. Meters and yards are units of measurement for length.

V

volume The amount of space occupied by a 3-dimensional object. A rectangular box that is 4 feet long, 3 feet wide, and 2 feet high has a volume of 24 cubic feet.

W

whole number A quantity that can be divided evenly by 1. For example, 3 and 14 and 2,500 are whole numbers, $\frac{2}{3}$ and $\frac{4}{5}$ are not whole numbers; they are fractions.

width Usually the measurement of the shorter side of a rectangle. A 3-inch by 5-inch photograph has a width of 3 inches.

Z

zero The mathematical symbol 0. When zero is added to any other number, the sum is that other number. For example, $5 + 0 = 5$.

INDEX